SpringerWienNewYork

CISM COURSES AND LECTURES

Series Editors:

The Rectors
Giulio Maier - Milan
Jean Salençon - Palaiseau
Wilhelm Schneider - Wien

The Secretary General
Bernhard Schrefler - Padua

Executive Editor
Carlo Tasso - Udine

The series presents lecture notes, monographs, edited works and proceedings in the field of Mechanics, Engineering, Computer Science and Applied Mathematics.
Purpose of the series is to make known in the international scientific and technical community results obtained in some of the activities organized by CISM, the International Centre for Mechanical Sciences.

INTERNATIONAL CENTRE FOR MECHANICAL SCIENCES

COURSES AND LECTURES - No. 479

ENVIRONMENTAL STRATIFIED FLOWS

EDITED BY

VINCENZO ARMENIO
UNIVERSITY OF TRIESTE, ITALY

SUTANU SARKAR
UNIVERSITY OF CALIFORNIA, USA

SpringerWienNewYork

The publication of this volume was co-sponsored and co-financed by the UNESCO Venice Office - Regional Bureau for Science in Europe (ROSTE) and its content corresponds to a CISM Advanced Course supported by the same UNESCO Regional Bureau.

This volume contains 133 illustrations

SPIN 11542353

In order to make this volume available as economically and as rapidly as possible the authors' typescripts have been reproduced in their original forms. This method unfortunately has its typographical limitations but it is hoped that they in no way distract the reader.

ISBN-10 3-211-28408-7 SpringerWienNewYork
ISBN-13 978-3-211-28408-7 SpringerWienNewYork

PREFACE

Stratified flows, common in environmental and geophysical applications, are characterized by the variation of fluid density in the vertical direction that can result in qualitative and quantitative modifications of the flow patterns by buoyancy . Unstable stratification (dense water/air above light water/air) increases the vertical mixing by generation of convective cells while stable stratification generally suppresses vertical mixing of mass and momentum. Even so, a stably stratified fluid can support internal waves, instabilities and turbulence that play a critical role in transport and mixing.

The ocean is predominantly subject to stable stratification which, under external excitation, supports an environment of internal waves which may then break and generate turbulence. Wind forcing, currents and convective plumes are other sources of turbulence in the ocean. In the ocean, stratified turbulence mediates the upward transport of bottom water, nutrients, chemical and biological species, and pollutants. In the atmosphere, stratification affects the transport of pollutants released at ground level, a critical problem being the thermal inversion in urban areas that causes the stagnation of pollutants and small particulate (PM2.5 to PM10) in the lower part of the atmospheric boundary layer. In buildings, stratification governs the circulation of air and heat in natural ventilation systems. Stratification plays a key role in determining the environmental and human consequences of accidents such as the release of gases into the atmosphere, fires in urban areas and wildland, and oil spills in the ocean. In industrial applications, the presence of a heavier phase that settles to the bottom in a mixture introduces stratification and consequent buoyancy effects on the flow. In hydraulic applications, a river that merges into the ocean basin tends to give rise to an interface between light fresh water sliding and spreading over the salty sea water so that the ensuing transport is very different from that in neutrally buoyant flow. The above-mentioned examples show that the analysis of stratified flows is of broad interest in geophysical, industrial and environmental applications.

The main objective of the current volume is to provide insight into the fluid mechanics of stratified flows with particular emphasis on turbulence and internal waves. The volume is designed for doctoral students as well as experienced researchers in engineering, geophysics, applied mathematics and physics, who are interested in broadening their knowledge in the area of stratified flows. The material spans basic concepts to recent advances in theoretical, numerical and experimental techniques. The volume is composed of four chapters. Chapter 1 by V. Armenio contains a brief description of turbulence together with some basic mathematical tools. The governing equations are introduced and recent algorithms for the numerical simulation of stratified flow are presented. Some insights on modeling stratified flows via large eddy simulation are given. Finally, a brief description of recent numerical results regarding stratified flows over topography are shown, in the contest of both atmospheric and ocean applications. Chapter 2 by C. Staquet deals with some basic notions on stably-stratified flows before focusing on the dynamics of internal gravity waves. The occurrence of stably-stratified flows in nature is illustrated and the Boussinesq approximation to the governing equations is derived. The properties of linearly-evolving waves are discussed together with the mechanisms that can lead the wave field to break. Parametric and buoyancy-induced instabilities, interaction with a shear flow, and interaction with a sloping boundary are also discussed. Finally, a briefly discussion about the statistical properties of the breaking wave field is given together with a brief introduction on mixing. Chapter 3 by S. Sarkar describes fundamental results on the properties of stratified turbulence and the associated mixing, obtained from direct numerical simulation (DNS) and large eddy simulation (LES) of building-block shear flows. The cases discussed include homogeneous turbulence evolving under the effect of linear stratification and linear shear, the shear layer, and channel flow. The chapter also deals with the novel situation where the mean shear is not aligned with the stratification, and the differences with the more classical case of alignment between shear and density gradient are highlighted. Finally, Chapter 4 by F. Nieuwstadt gives a review of the turbulent structure and dynamics of the atmospheric boundary layer. First, the equations governing the atmospheric boundary layer are given and discussed. The three main cases, namely the neutral, the stable, and the convective boundary layer are discussed in detail, and the appropriate scaling laws for each regime are identified. In addition, attention is paid to special topic such as wall roughness, the presence of vegetation, coherent structures, and clouds. The chapter concludes with a brief discussion of the non-stationary and inhomogeneous boundary layer evolving under different circumstances, from convective to stably stratified conditions.

During the final composition of the present volume, Prof. F. T. M. Nieuwstadt, author of chapter 4, passed away. He was a leader in fluid mechanics with a wide range of interests including turbulence, the atmospheric boundary layer and multiphase flows. He also was a very friendly and brilliant person and we enjoyed his presence during the summer school. We wish to dedicate the present volume to his memory.

Vincenzo Armenio and Sutanu Sarkar

CONTENTS

CONTENTS

Chapter 1:
Mathematical modeling of Stratified flows

Vincenzo Armenio

Dipartimento di Ingegneria Civile, Universitá di Trieste,
Piazzale Europa 1, 34127 Trieste, Italy

AbstractThe present Chapter of the lecture notes is divided into three different sections. The first section is devoted to the description of the equations governing a stratified flow field. The Reynolds averaged equations are derived together with the transport equations for the mean and turbulent kinetic energies. A background discussion on the spectral characteristics of a turbulent field is given, aimed at helping the comprehension of the successive sections. The second section describes the direct numerical simulation, together with the numerical techniques currently in use for the integration of the governing equations. Section 2 also contains a brief discussion on recent achievements of DNS in the study of stratified turbulent flows. Section 2 also deals with Large-eddy simulation of stratified turbulent flows. Models widely in use for the closure of the subgrid scale stresses are described and recent achievements in the field of stratified flows discussed. Section 3 is devoted to the description of very recent numerical results for stratified flows over a topography.

1 Governing equations for stratified turbulent flows

The numerical simulation of environmental stratified flows (ESF) is relevant in many applicative fields. As an example for weather predictions, for understanding of the biological cycle in the ocean or in lakes, for short-term prediction of dispersion of plumes and jets in large reservoirs. In the present section we deal with modeling of stratified flow and turbulence closures. In particular, very briefly we describe the techniques currently in use for large-scale predictions (usually based on the solution of the Reynolds-averaged governing equations), whereas, with more details we talk about the techniques employed for the description and understanding of small-scale processes. The general equations we are going to write work for a fluid dynamic field, and thus can be used for the simulation of the atmospheric boundary layer as well as for the simulation of water reservoirs (including oceans, lakes etc.).

The equations governing the dynamics of a stratified flow field are the well-known continuity and Navier-Stokes equations that rule the evolution of the velocity and pressure fields. The momentum equation contains a gravitational term that accounts for the effect of the density variation in the flow field. The evolution of the fluid density is considered through an advection-diffusion equation for the density field, or, equivalently, for the temperature and concentration fields that affect the fluid density by means of the equation of state. In Chapter 2 of the present lecture notes the Boussinesq approximation of the equations for a stratified flow is discussed. Herein we only intend to remind that the Boussinesq form of the equations holds under the following circumstances:

- The density variations (due to the variation of the temperature and concentration field) are small if compared to the bulk density of the fluid;
- The inertial accelerations are small when compared to the gravitational acceleration.

It is noteworthy that such approximations are valid in most environmental applications. Hereafter we consider a Cartesian frame of reference having the z (or x_3) axis oriented vertically upward) and the x,y axes (or x_1, x_2) contained in the horizontal plane. The velocity components will be u, v, w (or u_1, u_2, u_3) respectively along x, y, z. The Boussinesq form of the Navier-Stokes equations reads as:

$$\frac{\partial u_i}{\partial x_i} = 0, \tag{1.1}$$

$$\frac{\partial u_i}{\partial t} + \frac{\partial u_j u_i}{\partial x_j} = -\frac{1}{\rho_0}\frac{\partial p}{\partial x_i} + \nu\frac{\partial^2 u_i}{\partial x_j \partial x_j} - \frac{\rho}{\rho_0}g\delta_{i3} - f_i \times u_i \tag{1.2}$$

where f_i are the components of the rotation vector. As already mentioned, the density variation:

$$\rho(x, y, z, t) = \rho_T(x, y, z, t) - \rho_0$$

is related to changes of temperature and concentration of a dissolved phase in the fluid. In Eqs. 1.1,1.2, ρ denotes the perturbation density that varies in time and in space and ρ_0 is the constant, bulk density. The Boussinesq approximation requires $\rho/\rho_0 = \epsilon << 1$. In particular, for water reservoirs, the density variations are related to changes in the temperature field and in the concentration of dissolved salt (hereafter referred to as salinity). As regards the atmosphere, temperature and specific humidity need to be considered. A more detailed discussion is in Chapter 4 of the present lecture notes. The set of equation is completed once the perturbation density field ρ is evaluated.

It is possible to follow two different strategies: The first one consists in writing the transport equation of *all* quantities affecting the density field and successively to calculate the perturbation density as a function of the quantities already evaluated. This step is possible once the equation of state for the fluid is known.

For example, let's consider the case of the ocean. The density variation is related to corresponding variations of salinity (percentage of salt dissolved in a unit volume of water) and temperature. The equations of transport of the temperature and salinity are:

$$\frac{\partial T}{\partial t} + \frac{\partial u_j T}{\partial x_j} = \frac{\nu}{Pr}\frac{\partial^2 T}{\partial x_j \partial x_j}, \tag{1.3}$$

$$\frac{\partial S}{\partial t} + \frac{\partial u_j S}{\partial x_j} = \frac{\nu}{Sc}\frac{\partial^2 S}{\partial x_j \partial x_j}, \tag{1.4}$$

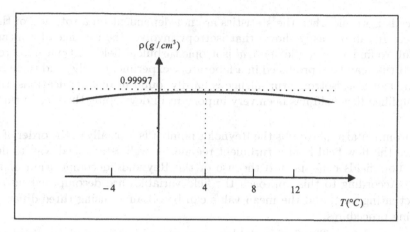

Figure 1. Variation of density of fresh water with temperature.

where T and S respectively denote the temperature and the salinity in the fluid, $Pr = \nu/k_T$, $Sc = \nu/k_c$ are the Prandtl and the Schmidt number. The Prandtl number is the ratio between the molecular diffusivity of momentum ν and that of temperature k_T, whereas the Schmidt number is the ratio between ν and the molecular diffusivity of the concentration (salinity) k_s. Now, the equation of state is needed that gives $\rho = f(T, S)$. For sea water the equation of state commonly used has been presented by Millero and Poisson (1981) and also given in the UNESCO Technical paper in Marine Science, Number 36 (UNESCO, 1981). A simplified (linearized) equation often used is the following:

$$\rho = \rho_0 + [-\alpha(T - T_0) + \beta(S - S_0)] \tag{1.5}$$

where α and β are respectively the coefficients of thermal and salinity expansion, and ρ_0 is the reference (bulk) density corresponding to the reference temperature T_0 and salinity S_0. Note that, in the range of values $T > 4^0C$, the increase of temperature gives a reduction of density (Fig.1), whereas the increase of salinity always produces the increase of density. Details on the complete equation of state for the ocean, for lakes, for dissolved gases in water and for the atmosphere can be found in Kantha and Clayson (2000).

1.1 Reynolds-averaged equations for stratified flows

Basic definitions that can be useful in the following are *homogeneity* and *isotropy*. Turbulence is homogeneous if the statistics of the turbulent field are not a function of space. Turbulence can be homogeneous over the whole volume of fluid, over planes or along a direction. As an example, a turbulent field characterized by the presence of a *linear* mean shear can be easily shown to be homogeneous over the volume of fluid, turbulence in a channel flow (the flow confined between two parallel and infinite plates) is homogeneous over the planes parallel to the plates, turbulence in a spatial-evolving boundary layer is homogeneous along the cross-stream, spanwise direction.

Turbulence is isotropic when the statistics are not dependent on a rotation of the frame of reference. It can be easily shown that isotropy involves the absence of a mean shear. It is difficult to find an example of a *real* isotropic turbulent field. A typical isotropic turbulent field that can be reproduced in a laboratory experiment is the grid turbulence. In spite of the fact that isotropic turbulence is hardly found in real applications, the study of this simplified flow field has been very important in developing theories in turbulence.

In environmental applications the Reynolds number is typically of the order of $O(10^7 - 10^9)$, hence the flow field is in a turbulent regime. A well established way to deal with turbulent flow fields consists into the use of the Reynolds decomposition of the field variables. According to this approach the field variables are decomposed into a mean and a fluctuating part, and the mean value can be obtained using three different types of averaging procedures:

1. Time averaging: The field variables are averaged in time and are function of the spatial coordinates

$$U_i(x_i) = \frac{1}{T} \int_0^T u(x_i, t) dt$$

2. Spatial averaging: The field variables are averaged along one direction (along which turbulence is homogeneous) or over planes or volume of homogeneity. If, for example, turbulence is homogeneous over the $x - y$ plane, spatial averaging leads to:

$$U_i(z, t) = \frac{1}{L_x L_y} \int_0^{L_x} \int_0^{L_y} u_i(x, y, z, t) \, dx \, dy$$

3. Ensemble averaging: It consists in averaging over a number of samples N of a particular experiment.

$$U^e(x_i, t) = \frac{1}{N} \sum_{n=1}^{N} u_i^n(x_i, t)$$

We assume that the reader is familiar with operations involving average and fluctuating quantities.

If we:

1. Substitute the total field variables with the sum of the mean and the fluctuating parts ($u_i = U_i + u_i'$) into the governing equations (continuity, momentum and transport equation of the temperature and/or concentration),

2. Perform an averaging operation, according to one of the techniques above described,

it is possible to obtain the following Reynolds averaged equations governing the evolution of the mean velocity pressure and density (or temperature or concentration) field.

$$\frac{\partial U_i}{\partial x_i} = 0 \tag{1.6}$$

$$\frac{\partial U_i}{\partial t} + \frac{\partial U_i U_j}{\partial x_j} = -\frac{1}{\rho_0}\frac{\partial p}{\partial x_i} + \nu\frac{\partial U_i}{\partial x_j \partial x_j} + \frac{g}{\rho_0}\overline{\rho} - \frac{\partial}{\partial x_j}\overline{u_i' u_j'} - f_i \times \overline{U}_i \tag{1.7}$$

$$\frac{\partial \overline{\rho}}{\partial t} + \frac{\partial \overline{\rho} U_j}{\partial x_j} = \frac{\nu}{Pr}\frac{\partial \overline{\rho}}{\partial x_j \partial x_j} - \frac{\partial}{\partial x_j}\overline{\rho' u_j'} \tag{1.8}$$

The Reynolds-averaged equations are formally similar to the governing equations for the instantaneous variables, apart the extra-terms on the right-hand side, that represent the contribution of turbulence to the mean velocity and density fields. The quantities $\rho_0\overline{u_i u_j}$ are the elements of the Reynolds stress tensor and represent transport of momentum due to turbulent fluctuations. Similarly, the quantities $\overline{\rho' u'}$ are transport of density by means of turbulent fluctuations, and they represent the contribution of the fluctuating field to the transport of the mean density. The momentum equation contains a gravitational term $\frac{1}{\rho_0}g\overline{\rho}$ that contributes to the enhancement or suppression of momentum depending on whether the flow is unstably or stably stratified.

Fluid flow involves conversion of energy from one form to another one. For example, kinetic energy is converted into heat by means of the molecular viscosity, consistently with the second principle of thermodynamics. When a gravitational term is present in the governing equations, due to the density variation in the flow field, conversion of kinetic energy into potential energy (or vice-versa) occurs in a stably (or unstably) stratified flow. In a stably stratified flow, for example kinetic energy is converted into potential one when a particle is displaced vertically in the fluid column. In order to understand how this conversion occurs, it is helpful to write the transport equation of the mean kinetic energy (MKE) K_M and of the turbulent kinetic energy (TKE) K:

$$K_M = \frac{1}{2}U_i U_i \,,$$

$$K = \frac{1}{2}\overline{u_i' u_i'} \,.$$

The transport equation for the mean kinetic energy is obtained multiplying respectively by U_j and U_i the momentum equation for U_i and U_j, and summing the two equations in order to obtain a transport equation for $U_i U_j$. If we put $i = j$ and we sum over the three directions we obtain:

$$\frac{\partial K_M}{\partial t} + \frac{\partial U_j K_M}{\partial x_j} = -\frac{1}{\rho_0}\frac{\partial U_j P}{\partial x_j} + \frac{\partial}{\partial x_j}[2\nu U_i S_{ij} - \overline{u_i' u_j'}U_i] + \overline{u_i' u_j'}S_{ij} - 2\nu S_{ij}S_{ij} - \frac{g}{\rho_0}i_3\overline{\rho}U_j \tag{1.9}$$

Equation 1.9 states that transport of mean kinetic energy in the flow field is due to the following contributions (the enumeration corresponds to the position of the terms of the right-hand side):

1. work done by the pressure forces (redistribution term);

2. transport by viscous forces and Reynolds stresses (redistribution term);
3. loss due to generation of turbulence (sink term);
4. energy dissipation due to viscosity (sink term);
5. work done by (or against) the gravitational field in unstably (stably) stratified flow. It represents a production (sink) term.

Terms 1 and 2 represent redistribution terms, in that they drop out when integrated over a closed volume; terms 3 to 5 are production-sink terms of MKE. Note that rotation of the frame of references does not contribute to production/destruction of MKE. Rather rotation leads to a redistribution of MKE among the three directions.

By the use of a similar procedure it is possible to derive the transport equation of the TKE:

$$\frac{\partial K}{\partial t} + \frac{\partial U_j K}{\partial x_j} = -\frac{1}{\rho_0}\frac{\partial \overline{u_j p}}{\partial x_i} - \frac{\partial}{\partial x_j}\left(\frac{1}{2}\overline{u_j u_i u_i}\right) + \frac{\partial}{\partial x_j}\left(2\nu\overline{u_j s_{ij}}\right) - \overline{u_i u_j}S_{ij} - 2\nu\overline{s_{ij}s_{ij}} - \frac{g}{\rho_0}\delta_{i3}\overline{u_j \rho'}$$

(1.10)

In Eq. 1.10 s_{ij} are the elements of the fluctuating strain-rate tensor. Equation 1.10 shows that turbulent kinetic energy is re-distributed by terms 1 to 3 of the right-hand side, whereas terms 4 to 6 represent production-sink terms of TKE.

In particular, term 4 of the RHS represents production of TKE. Note that $-\overline{u_i u_j}S_{ij}$ is in general positive and it appears in the equation of transport of MKE with the opposite sign. This means that what is subtracted from the mean flow for producing turbulence, appears in the TKE budget as a source term. Term 5 of the RHS of Eq. 1.10 is turbulent viscous dissipation whereas term 6 is a production (destruction) term in case of unstably (stably) stratified fluid column. In presence of the gravitational field, stratification in the fluid implies conversion between potential and kinetic energy. Figure 2a shows that, in case of unstable stratification vertical mixing of density is associated to negative values of the cross-correlation term $\overline{w'\rho'}$. In this case the buoyancy term 6 is positive, thus constituting a source term for TKE. Conversely, in case of stable stratification, the turbulent mixing of the density field is associated to a positive value of the cross-correlation term $\overline{w'\rho'}$ (Fig 2b); the buoyancy term 6 is in this case negative, thus acting as a sink term. To summarize:

- mean shear in the flow field always subtracts energy from the mean flow to sustain turbulence;
- the strain rate of the fluctuating field always gives rise to dissipation of turbulent kinetic energy; note that although the molecular viscosity is in general very small, the dissipation term $\epsilon = 2\nu\overline{s_{ij}s_{ij}}$ is such to be nearly in balance with the production term;
- stratification can give either production or destruction of TKE. In case of unstable stratification potential energy is converted into turbulent kinetic energy by means of the buoyancy term that acts as a source term in the TKE equation; In case of stable stratification, buoyancy term drains TKE that is thus converted into potential energy.

As already done for the kinetic energy, it is possible to derive a transport equation for the density variance. The procedure is similar to that already used for obtaining the transport equations for the Reynolds stresses. In particular:

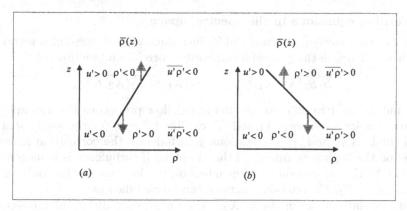

Figure 2. Schematic of turbulent density fluxes: (a) unstable stratification; (b) stable stratification.

- multiply the transport equation for the fluctuating density ρ' by ρ';
- Reynolds-average the terms of the equation.

We obtain the following equation:

$$\frac{\partial \overline{\rho'^2}}{\partial t} + \frac{\partial U_j \overline{\rho'^2}}{\partial x_j} = -\frac{\partial \overline{u_j \rho'^2}}{\partial x_j} - k_T \frac{\partial \overline{\rho'^2}}{\partial x_j \partial x_j} - 2\overline{u_j \rho'} \frac{\partial \overline{\rho}}{\partial x_j} - 2k_T \overline{\frac{\partial \rho'}{\partial x_j} \frac{\partial \rho'}{\partial x_j}} \qquad (1.11)$$

The first two terms of the RHS give transport by density fluctuations and viscous diffusion of the density variance. The third term gives production of density variance by means of mean density gradient, whereas the fourth term gives turbulence dissipation. It is worth noting that the production term $-2\overline{u_j \rho'}\, \partial \overline{\rho}/\partial x_j$ is always positive. Indeed, the cross-correlation $\overline{u_j \rho'}$ is negative when the density gradient is positive (unstable stratification), and $\overline{u_j \rho'}$ is positive in case of stable stratification (negative density gradient).

Equations 1.6, 1.7 and 1.8 involve mean (in the Reynolds-averaging sense) quantities and their spatial distribution also related to the effect of the turbulent field. The contribution of turbulence to the evolution of the velocity field is given by the elements of the Reynolds stress tensor $\rho_0 \overline{u_i u_j}$, whereas the contribution of turbulence to the evolution of the mean density field is given by the density fluxes $\overline{\rho' u_i}$. These elements are not known *a priori*, rather they represent additional terms that need to be determined (or modeled) in order to close the system of equations. There is no way to determine such terms analytically. For example, if we write the transport equations for the Reynolds stresses and the density fluxes, we obtain new unknown terms in the form of triple correlation terms. Similarly, the transport equation of the triple correlation terms contains $IV-$order terms that need to be determined. This is known in literature as the *closure problem*. Assumption are required for modeling the Reynolds stress and the density fluxes, with the aim to arrive to a closed system of equations. In the present notes we do not discuss the RANS turbulence models used in environmental and geophysical applications. For a detailed review of these techniques, the reader is referred to Kantha and Clayson (2000).

1.2   Governing equations in the spectral space

Herein, we give some definitions useful for introducing the concept of spectral analysis of turbulence. We define the general second-order correlation function as:

$$\overline{S_1 S_2}(X_1, X_2, t_1, t_2) = \overline{S_1(X_1, t_1) S_2(X_2, t_2)} \qquad (1.12)$$

where S_1 and S_2 are two different (or the same) flow properties (for example velocity components, density, pressure), X_1 and X_2 can be different (or the same) spatial locations, and similarly t_1 and t_2. For stationary turbulence, the correlation function does not depend on the beginning time t_1 of the sampling; if turbulence is homogeneous over the direction X, the correlation is independent on the location X_1 but only varies with the distance $X_1 - X_2$. Correlation functions can be classified as:

- Point correlations, when $X_1 = X_2$. They are called direct- or cross-correlation functions, respectively in the case $S_1 = S_2$ or $S_1 \neq S_2$;
- Spatial correlations, when $t_1 = t_2$. They are called direct-spatial correlations or spatial cross-correlations in the case $S_1 = S_2$ or $S_1 \neq S_2$ respectively.

Correlation functions are important in turbulence because they appear in the transport equations of momentum, kinetic energy and density as momentum fluxes $\overline{u_i' u_j'}$ (the elements of the Reynolds stress tensor) and density fluxes $\overline{\rho' u_i'}$. Indeed, let's consider the direct point correlation of a velocity component (say u_1), calculated for $t_1 = t_2$ for $X_1 = X_2$, $\overline{u_1 u_1}(X)$. This quantity is the longitudinal Reynolds stress divided by the reference density ρ_0. Similarly, the point cross-correlation between u_1 and u_3 for $t_1 = t_2$ and $X_1 = X_2$, $\overline{u_1 u_3}(X)$ is the Reynolds shear stress divided by ρ_0. Similarly for the other turbulent fluxes. If turbulence is homogeneous over the volume of fluid, the correlation is a function of the distance r between two points in the three-dimensional space, and the correlation tensor can be written as:

$$R_{ij} = \overline{u_i(X, t) u_j(X + r, t)}.$$

It is possible to define the spectrum tensor as the Fourier transform of R_{ij}, such that:

$$R_{ij}(r) = \int_{-\infty}^{\infty} \int_{-\infty}^{\infty} \int_{-\infty}^{\infty} F_{ij}(k) e^{i\,k\,r} dk$$

where k is the wave number, and the integral has been extended over the whole range of wavenumbers. Obviously the following inverse Fourier transform holds:

$$F_{ij}(k) = \frac{1}{(2\pi)^3} \int_{-\infty}^{\infty} \int_{-\infty}^{\infty} \int_{-\infty}^{\infty} R_{ij}(r) e^{-i\,k\,r} dr$$

This way it is possible to go forth and back from the physical space to the Fourier (or wavenumber) space. Based on the above discussion, it can be easily shown that:

$$2K = R_{11}(0) + R_{22}(0) + R_{33}(0) = \int_{-\infty}^{\infty} \int_{-\infty}^{\infty} \int_{-\infty}^{\infty} (F_{11} + F_{22} + F_{33}) dk$$

and, consequently the integral over the wave number of the sum of the spectra of the three velocity components is proportional to the turbulent kinetic energy. The function

$F_{ii}(k)$ (sum over i) gives the density of turbulent kinetic energy at the wavenumber k. Since $k_n = 2\pi/L_n$ with L_n a length-scale associated to the k_n-mode, the spectrum gives information on the energy content at different length-scales. If we consider a sphere of radius $k = |\overline{k}| = \sqrt{k_i k_i}$ and ds is the elementary surface of such sphere, the turbulent kinetic energy in the wavenumber space is:

$$E(k) = \frac{1}{2} \int_S F_{ii}(\overline{k}) ds,$$

where S is the surface of the sphere, and the turbulent kinetic energy is the integral of $E(k)$ over the whole range of wavenumbers.

Turbulence is characterized by the fact that energy is distributed over a wide range of scales (spatial and temporal). In particular, the shape of $E(k)$ informs on how energy is distributed over such scales. If we rewrite the transport equation of the TKE in the wavenumber space we can figure out the importance of the terms that contribute to the evolution of TKE over such scales.

Let's consider the equation of transport of kinetic energy (Eq. 1.10) and, for simplicity, let's assume the flow field to be homogeneous along the $x-y$-planes. This assumption is in general valid in the study of the atmospheric boundary layer, of the sea-bottom boundary layer and of the mixed layer in the upper part of the ocean. In case of homogeneity over the plane $x - y$ most terms of Eq. 1.10 drop out, and the equation simplifies to:

$$\frac{\partial K}{\partial t} = -\frac{\partial u_3' K}{\partial x_3} - \frac{\partial u_3' p}{\partial x_3} - \overline{u_1' u_3'}\frac{\partial U_1}{\partial x_3} + \frac{g}{\rho_0}\overline{\rho' u_3'} - \epsilon \tag{1.13}$$

where ϵ denotes the viscous dissipation. In order to understand how the terms that compose such equation are distributed over the different length-scales, it is convenient to recast such equation in the space of wave numbers. Such operation is quite complex and reported with details in Hinze (1976); Monin and Yaglom (1971). It gives:

$$\frac{\partial E(k)}{\partial t} = -\frac{\partial T_k(k,t)}{\partial k} + \tau(k)\frac{\partial U_1}{\partial x_3} + B(k) - 2\nu k^2 E(k) \tag{1.14}$$

In Eq. 1.14, $T_k(k,t)$ is the *spectral energy transfer rate* and represents the rate at which energy is transferred from a wave number k to $k + dk$ (from low to high modes). It has to be observed that the integral of $\partial T(k)/\partial k$ over the whole range of wave numbers gives 0. The second term of the RHS represents turbulent production of kinetic energy at the wave-number k, where $\tau(k)$ is the Fourier transform of the Reynolds stress tensor. The term $B(k)$ is the buoyancy production/destruction (depending on the sign of stratification) at the wave number k and, finally the last term is turbulent dissipation at the wave-number k. The following relationships hold:

$$\frac{g}{\rho_0}\overline{u_3 \rho'} = \int_0^\infty B(k)\, dk \tag{1.15}$$

$$P_k = \overline{u_1 u_3}\frac{\partial U_1}{\partial x_3} = \int_{-\infty}^\infty \tau(k)\frac{\partial U_1}{\partial x_3} dk \tag{1.16}$$

Figure 3 shows the distribution of the different terms of Eq. 1.14 over the wave numbers.

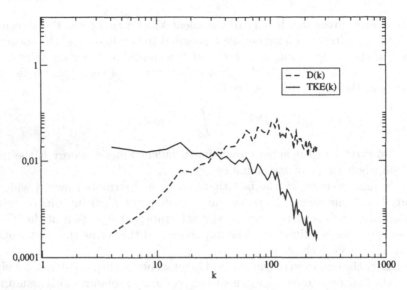

Figure 3. Distribution over the spatial wavenumbers of TKE and dissipation rate D of the turbulent kinetic energy. Data from a direct simulation of a turbulent channel flow at friction Reynolds number equal to 130.

It is worth observing that Fig. 3 refers to a case of large value of the Reynolds number. Turbulent kinetic energy is mainly concentrated in the range of low wavenumbers, corresponding to large length-scales. Hence, it clearly appears that the large scales carry most of turbulent kinetic energy. Dissipation is mainly concentrated in the region of large wavenumbers corresponding to small length-scales. Figure 3 also shows that the range of scales at which $E(k)$ is significant is well separated from that where dissipation occurs. Production of turbulent kinetic energy is significant at the small wavenumbers, since, such are the elements of the tensor $\tau(k)$. Finally, the intermediate range of scales over which both turbulence production and dissipation are negligible, is characterized by significant and nearly constant values of the spectral energy transfer rate. This range of scales, characterized by inviscid transfer of turbulent kinetic energy from the large to the small ones, is called *inertial subrange*. The classical theory of turbulence, corroborated by experimental evidence, states that the energy spectrum in the inertial range is proportional to $k^{-5/3}$. Finally, the buoyancy term (not shown in the Fig. 3) is significant at the small wave numbers, thus indicating that buoyancy production/destruction primarily affects the large, energy-carrying scales of motion.

The results discussed above confirm the theory of the universal equilibrium range postulated by Kolmogorov (1941b,a) on the basis of scaling arguments. The main points of such theory are:

1. At a Reynolds number large enough, there is a range of high wavenumbers where turbulence is statistically in equilibrium (production=dissipation) and determined by the dissipation rate ϵ and viscosity ν;

2. In the case of Reynolds number infinitely large, the energy spectrum at the intermediate scales (the above defined inertial subrange) is only dependent on the dissipation rate (ϵ) of TKE.

Based on the spectral analysis briefly described above, we try to give a short explanation and implications of the Kolmogorov theory. The largest scales of motion are responsible for the transport of momentum and generation of TKE. At these scales, the effective Reynolds number of the flow (based on the length of the largest scales and the characteristic velocity of these scales) is very large, and viscosity does not play a significant role. The non-linear terms of the Navier-Stokes equations prevent from the piling up of energy at the largest scales and are responsible of transferring energy from the largest scales toward the small scales. From a physical point of view this transfer occurs due to strong interaction and destruction of large-scale eddies into smaller and smaller eddies. This process would tend to go down in the length-scales (at infinitively large wave number), but there will be a certain length-scale η_k, characterized by a velocity scale v_k such that dissipative effects get dominant and supply complete dissipation of TKE. At these scales, the Reynolds number $Re_k = \eta_k v_k / \nu$ is expected to be of the order $O(1)$. Such scales are called the *Kolmogorov length- and velocity-scales*. The Kolmogorov theory assumes that the small-scales, characterized by small time scales, are statistically independent on the large and slow scales of motion. This hypothesis is confirmed by the analysis of the spectra of Fig. 3 that shows that the energy-carrying part of the spectrum is well separated by the part of the spectrum where dissipation takes place. When this assumption holds, the small scales of motion only depend on how they are fed by the energy coming down from the large scales and on the kinematic viscosity. Since in the intermediate-to-small scale turbulence is assumed to be in equilibrium, the rate of production of energy is equal to the rate of dissipation ϵ, and, hence, ϵ and ν are the key parameters that govern the smallest scales of the flow field. The application of the dimensional analysis suggests:

$$\eta_k \sim \left(\frac{\nu^3}{\epsilon}\right)^{1/4},$$

$$v_k \sim (\nu\epsilon)^{1/4},$$

$$t_k \sim \frac{\nu^{1/4}}{\epsilon}.$$

Note that the above expressions give $Re_k = \eta_k v_k / \nu \sim O(1)$. Finally, the existence of the inertial subrange (postulated by Kolmogorov and verified experimentally in successive researches) suggests that within this range energy is neither produced nor dissipated, hence the dissipation rate ϵ at the small scales can be directly related to the production rate at the largest scales. At these scales, the amount of kinetic energy per unit mass is proportional to u^2 where u is a characteristic velocity scale of the largest scales, and the rate of transfer of such energy is of the order of L/u where L is the characteristic length scale of the energy carrying eddies, it follows that the rate of energy supplied by the large, energy carrying scales to the small ones is of order u^3/L. Hence,

$$\epsilon \sim u^3/L$$

that implies that the large eddies lose a significant part of energy by non-linear mechanism within one turnover time u/L.

To summarize, for large-Reynolds-number flows:

- The spectral analysis allows to identify the energy content of the different scales of a turbulent flow field;
- Small wave numbers corresponding to large length-scales in the physical space carry most of the turbulent kinetic energy;
- Large wave numbers that correspond to small length-scales are characterized by being strongly dissipative;
- The large-scales where turbulent production takes place are well separated from the small scales by the inertial subrange where both production and dissipation are negligible and transfer from the large to the small scales takes place;
- the inertial subrange has universal characteristics such that $E(k) \sim k^{-5/3}$;
- Viscosity is unimportant in the dynamics of the large scales. On the other hand, the small scales are governed by the viscosity and the dissipation rate of TKE;
- The dissipation rate can be found by the knowledge of the dynamics of the large, energy-carrying scales of motion.

Similar arguments can be used for the analysis of the density field.

Let's consider the transport equation of the density variance. In particular, we discuss the case of a passive scalar, namely a temperature or concentration field weak enough that the feed-back effect on the momentum equation is negligible. The passive scalar represents a case one-way coupling, in which the scalar is transported by the velocity field, and it does not affect the velocity field. A boundary layer heated from one boundary (either the top or the bottom), where the heating is very weak and not able to supply significant buoyancy, is a typical example of transport of a passive scalar. The transport equations of the mean passive scalar and of its variance are respectively Eq. 1.8 and Eq. 1.11 where $\bar{\rho}$ and ρ' have to be considered as mean and fluctuating parts of the passive scalar. Similarly to the case of energy spectra, the passive scalar has a three-dimensional spectrum. Let's consider a density fluctuation $\rho'(X, t)$ and the spatial autocorrelation function R_ρ defined as:

$$R_\rho = \overline{\rho(X, t)\rho(X + r, t)},$$

where r is the distance in the three-dimensional space from the point X to the point $X + r$. The Fourier transform of R_ρ is the spatial spectrum $F_\rho(k)$ such that:

$$R_\rho(r) = \int_{-\infty}^{\infty} \int_{-\infty}^{\infty} \int_{-\infty}^{\infty} e^{ikr} F_\rho(k) \, dk.$$

The spatial spectrum $F_\rho(k)$ is obviously defined as:

$$F_\rho(k) = \frac{1}{(2\pi)^3} \int_{-\infty}^{\infty} \int_{-\infty}^{\infty} \int_{-\infty}^{\infty} e^{-ikr} R_\rho(r) \, dr.$$

Similarly to the velocity field, the density spectrum represents the *energy* of the density field associated to the wavenumber k. If we integrate over the surface S of the sphere of radius k the function $F_\rho(k)$, we obtain the three-dimensional spectrum:

$$E_\rho(k) = \frac{1}{2} \int_S F_\rho(\overline{k}) \, ds,$$

where $k = |k| = \sqrt{k_i k_i}$ and ds is the elementary surface of the sphere of radius k. Obviously, the density variance is related to the three-dimensional spectrum by:

$$\frac{1}{2} \overline{\rho' \rho'} = \int_0^\infty E_\rho(k) \, dk.$$

An important point is that if an equilibrium range exists in the spectrum of the turbulent kinetic energy, there exists an equilibrium range in the spectrum of the density variance. This is reasonable since turbulence is responsible of the mixing of the scalar field and, thus the spectrum of the scalar field must have characteristics somewhat similar to those of the velocity field. The existence of an inertial subrange in the spectrum of the passive scalar has been first argued by Obukhov (1949) and Corrsin (1951). Similarly to the energy spectrum, such subrange (called inertial-convective subrange) is such that variance of passive scalar is neither produced nor dissipated, rather it is transferred from the large-scales to the small-scales of motion. The equation of transport of the density variance in spectral space gives terms equivalent to those of Eq. 1.14 . A production term is present that basically is fed by the large scales of motion and therefore is significant in the region of small wavenumbers. A dissipation term ϵ_ρ appears that is relevant in the range of small scales of motion, corresponding to high wavenumbers, and a transfer function appears that is significant at the intermediate scales, where both production and dissipation are negligible. Such inertial-convective subrange is present at intermediate wavenumbers in the following range:

$$\frac{1}{l_\rho} << k << \frac{1}{\eta_\rho},$$

where l_ρ is an integral scale at which the variance of passive scalar is produced, and η_ρ is the dissipation micro-scale, the smallest scale detectable in the spectrum. This scale is the counterpart of the Kolmogorov micro-scale of the energy spectrum, and it is called the Obukhov scale. Similarly to the inertial subrange, the inertial-convective subrange has been found to decrease with $k^{-5/3}$. Figure 4 shows the spectrum of passive scalar as derived from experimental data of Sreenivasan (1996) for cases characterized by large values of the Reynolds numbers.

A main difference between the spectrum of the velocity field and that of the density variance is that the latter is dependent on an additional parameter that is the Prandtl number (and/or the Schmidt number), namely the ratio between the fluid viscosity and the thermal (or concentration) diffusivity.

Dimensional arguments (see Tennekes and Lumley (1972)) suggest:

$$E_\rho = \epsilon_\rho \epsilon^{-1/3} k^{-5/3} f(k\eta, Pr)$$

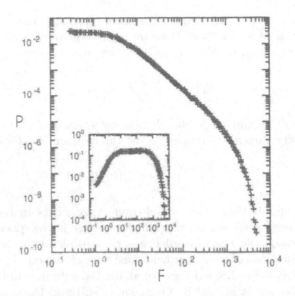

Figure 4. Spectrum of passive scalar obtained by experimental data. Reproduced with permission from Sreenivasan, (1996).

It is noteworthy that the shape of the spectrum depends on the characteristics of the fluid, due to the presence of Pr. If the Reynolds number is large enough for the development of an inertial subrange, and the thermal diffusivity is small enough, there is a large part of the spectrum where thermal diffusivity is unimportant and an inertial-convective range is detectable. In this case the spectrum of this range is:

$$E_\rho = \beta \epsilon_\rho \epsilon^{-1/3} k^{-5/3}$$

Recent measurements of Sreenivasan (1996) show values of $\beta \sim 0.4 - 0.5$. This is indeed the case of a passive scalar like salinity, characterized by $Sc \sim 500$.

When the thermal diffusivity is to be taken into account, we must consider separately fluids with $Pr < 1$ (like diffusion of temperature in air) and fluids with $Pr > 1$ like , for example, temperature diffusion in water ($Pr \sim 5$). The case $Pr < 1$ is such that the thermal diffusivity gets importance within the inertial-convective range where viscosity is still unimportant. The Obhukov scale is larger than the Kolmogorov scale and the spectrum of density variance decays much faster than the $k^{-5/3}$ law. The region of wavenumbers $1/\eta_\rho < k < 1/\eta_k$ is called the *inertial-diffusive region*.

The case $Pr > 1$ is characterized by the fact that viscosity becomes important in a region of the spectrum where the effect of thermal diffusivity is still negligible. The range of wavenumbers where $k\eta > 1$ and $k\eta_\rho << 1$ is called *viscous-convective* subrange and here $E_\rho \sim k^{-1}$. In this case there is also a region where $k\eta_k >> k\eta_\rho > 1$ called *viscous diffusive* range. The ranges above illustrated are shown in Fig. 5

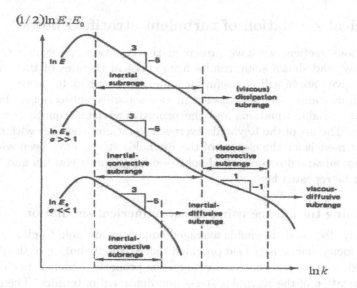

Figure 5. Examples of spectra of passive scalar for different values of the Prantl number. Reproduced with permission from Tennekes and Lumley, (1972).

When buoyancy effects become important the spectrum turbulent kinetic energy an that of the density variance exhibit different characteristics. This aspect is exploited in Chapter 2 of the present lecture notes.

2 Numerical simulation of turbulent stratified flows

In the previous section, we have presented the equations governing the motion of a stratified flow, and shown some results from classical theories of turbulence. The few information given above will be useful for understanding how to tackle turbulence using numerical simulations. We will discuss in two separate subsections, the direct use of the primitive variable equations and the primitive variable equations with a subgrid-scale closure. The use of the Reynolds-averaged equations, together with description and discussion of models for the closure of the Reynolds stress, have been widely treated in other courses, advanced schools and books (see for example Kantha and Clayson (2000) and will not be repeated here.

2.1 Modeling turbulence using direct numerical simulation

As already discussed, Reynolds averaged equations are able to give an estimation of the mean velocity and density field provided that the contribution of the fluctuating field (the Reynolds stresses and the density fluxes) be properly modeled. Such model supplies a parameterization of the Reynolds stress and density flux tensors. The main drawback of turbulence models for Reynolds averaged equations is in the dependence of the results on a series of empirical parameters to be tuned case by case. On the other hand, the advantage of using RANS, consists in the possibility to deal with values of Reynolds number of practical significance.

On the opposite side, one may be tempted to simulate a turbulent flow, solving *all* the scales of turbulence directly, by direct simulation of the NSEs without the use of any turbulent closure. Such an approach is called *direct numerical simulation* (DNS). In order to understand how this operation can be performed, and the limits of such simulations, we need to go back and discuss some issues on the characteristics of a turbulent flow field.

As discussed in a previous section, turbulence is characterized by a cascade of energy from the large, energetic scales toward the small ones. The large scales are energetic and anisotropic. The evolution of such scales is strongly related to the boundary condition on the flow field, and hence their structure is strongly dependent on the physical problem under investigation. As an example the shape and the temporal evolution of the spatial large scales in a boundary layer are very different from those developing in a shear layer, or in a turbulent jet. Conversely, the small scales of turbulence tend to be more universal, in that they tend to be more and more isotropic and strongly dissipative. Figure 6 shows a typical energy spectrum for a large class of neutrally stratified turbulent flows (Saddoughi and Veeravalli (1994)). Figure 6 contains the spectra calculated from data obtained for a wide range of values of the Reynolds number and for different kinds of experiments (for example, grid turbulence, shear flows, boundary layers...). The energetic part of the spectrum is observable at the left side of the spectrum, corresponding to small wavenumbers (large spatial scales). In some of the experiments (those carried out at large Reynolds numbers) the presence of the inertial subrange is well detectable in the range of moderate wavenumbers. Finally, the large wavenumbers are characterized by fast decay of energy spectrum. The analysis of Fig. 6 also shows that:

- different flow conditions are characterized by different levels of energy content at

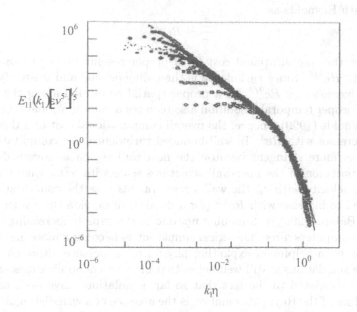

Figure 6. Re-scaled spectra of a wide variety of turbulent flows. Reproduced with permission from Saddoughi and Veeravalli, (1994).

the small wavenumbers;

- the inertial range and the small-scale part are characterized by the collapse of the rescaled spectra obtained in different flow field (as also postulated by the *equilibrium range theory*);
- the width of the inertial range over the wavenumbers increases with the Reynolds number. In other words, the distance (in the domain of wavenumbers) between the scales where turbulence is produced and those over which turbulence is dissipated increases with the Reynolds number.

It has to be remarked that all the implications of the *equilibrium range theory* are based on a key hypotheses, namely the Reynolds number is assumed to be very large such that the scales where turbulence production takes place are well separated from those where dissipation occurs.

A direct simulation, by definition, must be able to simulate all the scales of motion. In particular the computational domain must be large enough to reproduce correctly the space-time evolution of the largest and energetic scales, that, as already discussed are governed by the boundary conditions. The grid size must be small enough to capture the smallest scales where energy dissipation occurs (Kolmogorov scales). In the previous section, the dissipation rate has been shown to be proportional to the large scales of turbulence $\epsilon \sim u^3/L$ where L is a characteristic length-scale of the large eddies of the flow field. Since the Kolmogorov scale can be estimated from the knowledge of ϵ and ν, it can be easily shown that the ratio between the largest scale L and the smallest scale

η increases with Reynolds as:

$$\frac{L}{\eta} \sim Re^{3/4}$$

It follows that the computational cost for a proper resolution *in just one direction* is proportional to $Re^{3/4}$. Since turbulence is three-dimensional and unsteady, the computational cost increases as $Re^{9/4}$ for a proper spatial resolution along the three spatial directions. A proper temporal resolution also requires a computational cost proportional to $Re^{3/4}$ (Reynolds (1990)), hence, the overall computational cost of a direct numerical simulation increases with Re^3. In wall-bounded turbulence, the computational requirements are even more stringent because the near-wall region is governed by viscosity. Indeed, the dimension of the near-wall structures scale with ν/u_τ where $u_\tau = \sqrt{\tau_w/\rho_0}$ is the friction velocity with τ_w the wall stress. In this case the computational cost has been estimated to increase with $Re^{3.5}$ (for a detailed discussion the reader is referred to Piomelli and Balaras (2002)). Summing up, due to the rapidly increasing capabilities of modern supercomputers direct numerical simulation is becoming more and more attracting. DNS has been helpful in exploiting physics of turbulence although the Reynolds number of the simulations is still well below that of practical applications. A main drawback of DNS, associated to the fact that so far, simulations have been carried out for very small values of the Reynolds number, is the absence of a well distinguishable inertial range. Considering that most well established theories have been developed in the limit $Re \to \infty$, characterized by an infinite extension of the inertial subrange, it is still to be clarified how results coming from DNS can be extrapolated in the range of large Reynolds numbers.

Numerical techniques for DNS   Direct numerical simulation is nowadays considered an effective tool for understanding physics of turbulence. In this section we briefly discuss numerical techniques widely used for the direct solution of the Navier-Stokes equations. First, it has to be pointed out that, since a turbulent flow field is always three-dimensional and unsteady a DNS must have these characteristics, and hence requires the use of unsteady, 3D solvers.

As regards the time-advancement of the NSEs, the use of a numerical technique designed for unsteady-flow calculations is required. Due to its own conservation properties, the fractional step technique is currently in use for such simulations. A basic, second-order accurate-in-time algorithm is herein described (for details see Armenio (1997)). The momentum equation is written as:

$$\frac{u_i^* - u_i^n}{\Delta t} = -\frac{1}{2}[3(C(u_i) + \frac{g}{\rho_0}\rho_p)^n - (C(u_i) + \frac{g}{\rho_0}\rho_p)^{n-1}] + \frac{1}{2\,Re}\nabla^2(u_i^* + u_i^n) \quad (2.1)$$

$$\frac{u_i^{n+1} - u_i^*}{\Delta t} = -\nabla\Phi^{n+1} \quad (2.2)$$

where C denotes the advective, non-linear term, u_i^* is an intermediate, unphysical velocity field, n and $n+1$ respectively denote the time instants t and $t + \Delta t$ and Φ is a computational pressure field that projects the velocity field into a divergence-free one. If one takes the divergence of Eq. 2.2 and forces the velocity field to be solenoidal at the $n + 1$ time instant, the following Poisson equation is obtained:

$$\nabla^2 \Phi^{n+1} = \frac{1}{\Delta t} \nabla \mathbf{u}^* \tag{2.3}$$

Kim and Moin (1985) have shown that, in the semi-implicit formulation of Eq. 2.1 the physical pressure is related to the operator Φ by:

$$p^{n+1} = \Phi^{n+1} + \frac{\Delta t}{2Re} \nabla^2 \Phi^{n+1} \tag{2.4}$$

The density equation is discretized similarly to 2.1 as follows:

$$\frac{\rho_p^{n+1} - \rho_p^n}{\Delta t} = -\frac{1}{2}[3C(\rho_p)^n - C(\rho_p)^{n-1}] + \frac{1}{2\,Re\,Pr}\nabla^2(\rho_p^{n+1} + \rho_p^n). \tag{2.5}$$

The algorithm described above, belongs to the more general class of fractional-step techniques. In Eqs 2.1 2.5 the advective terms and the explicit forcing term (in Eq. 2.1 only) are integrated explicitly by the use of the Adams-Bashforth techniques, whereas, the diffusive terms are treated implicitly by means of the Crank-Nicolson scheme. The use of an implicit technique allows avoiding the diffusive limitation on the time step. Originally, this technique has been designed for Cartesian regular meshes, although more recently it has been re-formulated for equations written in curvilinear coordinates (see Zang et al. (1994)). The latter has allowed to deal with problems characterized by moderately complex geometry (see for example Zang and Street (1995); Armenio and Piomelli (2000); Calhoun and Street (2001); Calhoun et al. (2001); Ding et al. (2003); Falcomer and Armenio (2002); Armenio et al. (2003) both for neutrally and for stably stratified turbulent flows. A modified fractional-step, that uses third- or fourth-order Runge-Kutta schemes for the time-advancement of the advective terms is also commonly in use (details are in Rai and Moin (1991)). It is noteworthy that such procedure, when implemented in conjunction with the Crank-Nicolson scheme for the diffusive term, produces an algorithm that is accurate at the second-order in time. The only advantage of using the RK3 or RK4 time-advancement technique is that the advective time-step constraint is now less severe. In particular the $2nd$-order Adams-Bashfort scheme requires the following restriction on the Courant-Friedrichs-Levi (CFL) number:

$$CFL = \frac{u_1\Delta t}{\Delta x} + \frac{u_2\Delta t}{\Delta y} + \frac{u_3\Delta t}{\Delta z} < 1,$$

whereas, the RK4 scheme requires $CFL < \sqrt{3}$. Recently a simple fractional step algorithm has been developed by Slinn and Riley (1998) for dealing with turbulent stratified flows. This algorithm is fully explicit and the terms of the momentum equations are treated using the $3rd$-order Adams-Bashforth technique:

$$\frac{u_i^* - u_i^n}{\Delta t} = \frac{23}{12}TERM^n - \frac{16}{12}TERM^{n-1} + \frac{5}{12}TERM^{n-2} \tag{2.6}$$

where TERM includes the advective the diffusive and the gravitational terms. The other equations of the fractional step method remain unaltered. A similar equation can be written for the transport of the perturbation density. The stability limits of the present algorithm are two, namely an advective condition and a diffusive one. The former is:

$$\sqrt{3}CFL + \frac{g}{\rho_0}\rho_p\Delta t < 0.724$$

The latter is:

$$\frac{\nu\Delta t}{\Delta x^2} + \frac{\nu\Delta t}{\Delta y^2} + \frac{\nu\Delta t}{\Delta z^2} < 0.5$$

Note that the semi-implicit algorithm did not have any diffusive limit, due to the implicit treatment of the diffusive term. In direct or large-eddy simulations (the topic of the next section), the convective limit is usually more stringent than the diffusive one, due to the very small values of the kinematic viscosity and scalar diffusivity. This holds in *resolved* LES as well (whose definition will be given later within the present Chapter), since in these cases the eddy viscosity (or diffusivity) are of the order of the 50%-60% of the molecular one. A main advantage of the explicit algorithm compared to the semi-implicit one, stands in its own scalability when implemented for MPI parallel supercomputers. This is an important point when simulations involving tents of millions of grid cells are carried out. Finally, the application of the explicit algorithm of Slinn and Riley (1998) to the curvilinear formulation of the NSEs, using the technique of Zang et al. (1994) appears to be straighforward.

Spatial discretization plays an important role in the overall accuracy of the numerical scheme. Either spectral methods or finite difference methods are employed for the spatial discretization of the derivatives. In order to understand the performance and the limitations of these schemes we can consider the very simple function $f(x) = e^{ikx}$ which has analytical derivative $f'(x) = ike^{ikx}$. If we differentiate the function f numerically, we obtain the following function:

$$\frac{\delta f}{\delta x} = ik'e^{ikx}$$

where k' is the modified wavenumber. The modified wavenumber is complex, the real part represents the attenuation of the numerical derivative compared to the analytical one whereas the imaginary part represents the phase error introduced by the numerical scheme. Figure 7 shows the real part of the wavenumber obtained using different kinds of spatial discretization. Spectral methods appears very accurate, due to their own ability to reproduce the whole range of wavenumbers. On the contrary, *2nd* order, centered finite difference schemes give good resolution of the small wavenumbers and poor resolution of the large wavenumbers. High order finite difference schemes give intermediate resolution between the *2nd*-order and the spectral schemes. The high order schemes are computationally much more expensive than the *2nd*-order centered scheme. The need to solve the large wavenumbers, that correspond to the small-scale fluctuations, requires one of the following strategies:
- to use a high-accuracy scheme on a certain grid size;
- to use low-order schemes (accurate at least to the *2nd*-order) increasing the grid resolution.

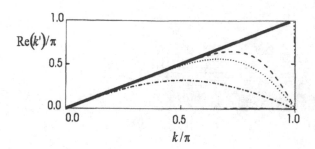

Figure 7. Real part of the modified wavenumber for different spatial-discretization schemes: solid line exact; dot-dot-dot-dashed line, spectral scheme; dashed line, fourth-order Padé; dotted line, fourth-order compact scheme; dot-dashed line, second order centered scheme. Reproduced with permission from Piomelli and Chasnov (1996).

A choice of the numerical scheme to be used must take into account the flexibility of the method in the application of the boundary conditions. When periodicity can be used in one or more directions (due for example to the fact that turbulence is homogeneous along these directions) spectral schemes are well suited. On the other hand, for inhomogeneous flows, (spatially evolving boundary layer or mixing layers) 2nd-order central finite differences offer more flexibility in handling the inflow-outflow boundary conditions. Kim et al. (1987) have used a spectral method for their keystone direct simulation of the turbulent field developing between two infinite and parallel plates (channel flow) using Fourier expansion in the planes of homogeneity (parallel to the walls) and Chebychev polynomials in the inhomogeneous, wall-normal, direction. Mixed techniques are also used whenever convenient. For example, Garg et al. (1997) have studied the stably stratified flow in a closed and open channel (homogeneous along the horizontal planes) using Fourier expansion in the horizontal direction and finite differences along the wall-normal direction.

A drawback associated to the use of spectral methods or 4th-order schemes is that they are not designed for working in curvilinear coordinates. The use of the curvilinear formulation of the Navier-Stokes equations allows simulating turbulent flows over moderately complex geometry. The use of such formulation implies the transformation of the physical (irregular) domain into a three-dimensional, rectangular and uniform computational box. Such transformation is carried out through the metric tensors, whose elements must be evaluated numerically. In order to achieve a certain overall spatial accuracy, (see Thompson et al. (1985)) the same numerical scheme is required for the evaluation of the space derivatives and of the metric terms. This poses a severe limitation on the application of high-order schemes to the analysis of the flow fields over irregular geometry, since a consistent implementation would require the use of high order

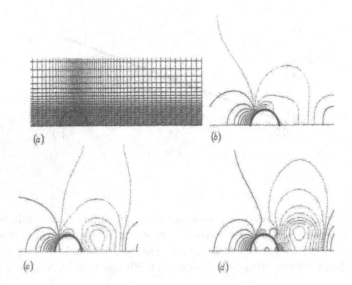

Figure 8. Computational grid for the investigation of the flow around a sphere and iso-lines of pressure field. Reproduced with permission from Fadlun et al., (2000).

discretization formulas for the computation of all the metric terms.

An alternative strategy for dealing with complex geometry, still holding the advantages of using Cartesian geometry is the use of the recently developed *immersed boundary* technique Fadlun et al. (2000). According to such technique,boundary body forces allow the imposition of the boundary conditions on a given surface not coinciding with the computational grid (Fig. 8). Since in general the surface of the body intersects the grid cells, proper interpolation techniques are required to calculate the velocity at the boundary. In spite of its own flexibility in dealing with very complex geometry (combustion chambers etc.,) a drawback of the immersed boundary technique is that numerical noise can be generated near the boundary cells, due to loss of precision in the interpolation of the velocity field at the body surface.

The proper simulation of a turbulent (neutrally or stably stratified) field needs careful use of the boundary conditions. As above discussed, periodic conditions are well suited for flow fields that exhibit homogeneity along a direction. The use of such condition implies that the flow field repeats identically along the direction of periodicity and, thus poses a limitation on the largest scale of the flow that can be simulated. It has been shown (see Moin and Mahesh (1998)) that a domain is large enough to simulate the significant large scales of the flow when the two-point correlation of the velocity field decays within half domain in the direction of homogeneity. Typical cases in which periodic conditions have been successfully used are:

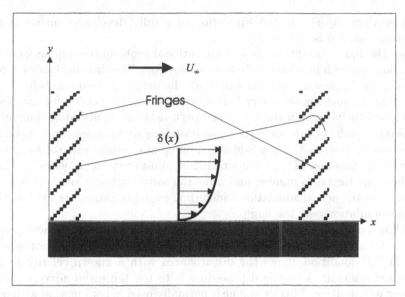

Figure 9. Schematic of the fringe regions for the numerical simulation of a boundary layer.

- forced and/or decaying isotropic turbulence, homogeneous turbulence (forced by a linear and steady mean shear). In these cases triple-periodic box are commonly employed;
- open or closed channel flow. Periodicity is usually imposed over the planes of homogeneity;
- spatial evolving boundary layer. Periodicity is imposed over the spanwise direction of homogeneity;
- flow over transversal two-dimensional wavy-walls, backward facing step. Periodicity is imposed over the spanwise direction of homogeneity;
- time evolving mixing-layers or shear-layers. Periodicity imposed over the planes orthogonal to the direction of the shear.

In order to take advantage of the simplicity of periodic boundary conditions, Spalart and Wattmuff (1993) have developed a *fringe method* that allows using periodicity in a spatially evolving boundary layer. According to such technique, additional terms are added at the upstream and downstream parts of the computational domain, whereas the standard NSEs are solved in the remaining. These terms act toward a reduction of the boundary layer thickness at the right edge of the domain with the aim to make the flow periodic along the streamwise direction (see Fig. 9).

Problems characterized by grid topology that changes along the streamwise direction cannot be treated by the use of periodic conditions. In this case the imposition of inflow-outflow conditions is required. The inflow condition can be easily imposed in transitional fields (for example one can impose the Blasius profile of the zero pressure gradient boundary layer). In the simulation of a fully developed turbulent field two different strategies can be followed:

- to use the inflow conditions as in a transitional problem in conjunction with a domain long enough to allows turbulence to develop. The drawback of such technique is that even in presence of large-amplitude disturbances, that can help in triggering transition to turbulence, a very long upward domain is needed for reaching a fully developed turbulence in the downward part of the computational domain;
- to use the results of a periodic *pre*-simulation for a non-periodic one. As an example, if one has to simulate the flow field over a bump at a certain value of Re, it is possible to create a database of time-evolving inflow planes running a periodic channel flow at the same Reynolds number and using the same cross-stream distribution of grid cells, as in the actual simulation. Such time-evolving planes are used at the inflow of the simulation over the bump.

As outflow conditions, the best one can hope is that the disturbances produced at the outflow do not propagate upward deteriorating the accuracy of the simulation. The Orlansky (1976) condition allows the disturbances with a known celerity to leave the computational domain. A standard procedure is to use the mean advective velocity of the flow field as celerity c. This technique is normally used in conjunction with other ones that help in diffusing the disturbances reflected back from the outflow section. Among them, the smooth, artificial increase of viscosity by more than one order of magnitude in the outflow region (for instance, by the use of a cubic or exponential law) appears very simple and effective (Verzicco, *private communication*).

Many problems in stratified turbulent flows are characterized by reflection of internal waves within the physical domain. This is indeed the case of stratified flows over a topography under tidal motion. In these cases, a problem arises about the boundary condition to be used for the perturbation density. Orlansky-like conditions are not suggested, due to the fact that these conditions are suited for problems characterized by the fact that the phase speed and the group-speed are aligned. On the other hand in internal-wave problems, the phase-speed and the group-speed are orthogonal to each-other. This means that if a BC allows the wave to leave the domain, for example through a wave equation with celerity equal to the phase speed, energy remains confined within the domain due to the fact that the group velocity is parallel to the boundary. For such problems, Slinn and Riley (1998) suggest the use of a sponge region to be located in the region where spurious reflection of internal waves is expected. A Rayleigh damping of the form:

$$\bar{\theta} = \theta_i - \sigma_i(\theta_i - \theta_{i0}) \tag{2.7}$$

has been shown to work properly. In Eq. 2.7 $\bar{\theta}$ is the damped value of a field variable (for instance u_i or ρ_p), θ is the not damped value of the function and θ_{i0} is the expected value of the function at the boundary. The function σ_i is a damping coefficient. A form of this coefficient that has been found to work satisfactorily is the following Gaussian function:

$$\sigma_3 = A \, e^{-(3.5z_d/L_d)^2/2}, \; for \; 0 \leq z_d \leq L_d$$

where A is an empirical damping coefficient and $z_d = z - (L_z - L_d)$ with L_z and L_d respectively the height of the domain and the width of the sponge region. The above

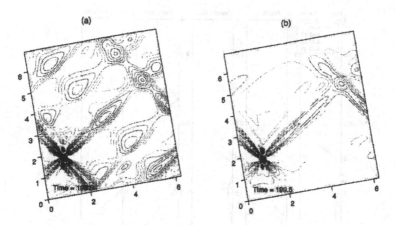

Figure 10. Contour lines of perturbation density within a computational box. Internal waves are generated at a source point. a) simulation without the application of the sponge regions; b) simulation with the sponge regions. From Taylor et al. (2004b).

described sponge region is best suited for damping short waves, whereas it appears less effective in damping large sub-harmonic waves that can be generated within the computational domain. In order to prevent from continuity errors, Slinn and Riley (1998) have implemented the damping function before the projection step as follows:

$$u_i^{n+1} = u_i^* - \sigma_i(u_i^* - u_{i0}) - \Delta t \frac{\partial p^{n+1}}{\partial x_3}$$

letting the sponge region to work in the vertical direction only. Very recently, Taylor et al. (2004b) have tested such non-reflective BCs and have found that good performances can be obtained if the thickness of the sponge region is large enough. Figure 10a shows the results of a test performed generating internal waves within the computational box. The perturbation density is constant at the boundaries. Large reflections occur at the top-bottom boundaries that, after a while, deteriorate the numerical solution. On the other hand, Fig. 10b, shows that with a proper choice of the thickness of the sponge region, choice that need to be made empirically, good results are obtained. Figure 11 shows that the variation of the damping coefficient allows to reduce the energy content of the reflected waves at the fundamental frequency as well as at the super harmonics. On the other hand, the sub-harmonics generated during the spurious reflection process are much less affected by the damping function.

As far as initial conditions are concerned, they are unimportant, since, one is in general interested in the simulation of statistically steady turbulence. In order to reduce the time for reaching such a condition, a good choice is starting from a fully developed turbulent field, for example interpolated from one available in the database of the research group. The simulation must be run up to the achievement of a statistically steady condition. Finally, the actual simulation can be run and the statistics of the turbulent field accumulated (Fig. 12). Statistical convergence is usually obtained within 8 − 10

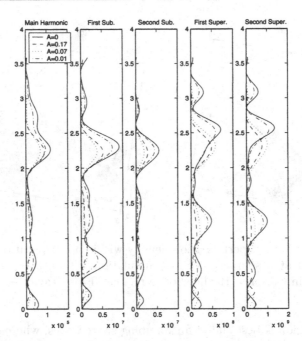

Figure 11. Energy content at different wavenumbers at a vertical cut of the domain of Fig. 10 for different values of the damping coefficients. The plots go from the left to the right showing the energy content from the sub-harmonics to the super-harmonics: $2nd$ sub-h; $1st$ sub-h; forcing frequency; $1st$ super-h; $2nd$ super-h. From Taylor et al. (2004b).

eddy turnover times.

In cases characterized by the transport of a scalar (either passive or active), it has to be considered whether the grid resolution is enough for reproducing the scales at which dissipation of the scalar takes place. Referring to the case of a passive scalar, the examples of Fig. 5 clarify this concept. In cases characterized by $\eta_\rho > \eta_k$, and thus for values of the Prandtl number (or equivalently of the Schmidt number) smaller than 1, a grid resolution suited for simulating the whole spectrum of kinetic energy is enough for the direct simulation of the scalar. This is typically the case of the evolution of a temperature field in air ($Pr = 0.71$). Conversely, Fig. 5 shows that for $Pr > 1$ the grid size for solving the small scales of the scalar field has to be much smaller than that required for the evaluation of the spectrum of kinetic energy. Typically, this is the case of the evolution of a temperature field ($Pr \sim 5 - 7$) or salinity ($Pr \sim 500$) in water.

Finally, when the scalar is active (in case of stratified flows), the scale of dissipation of the scalar field changes compared to the neutrally stratified flow. For instance, stable stratification reduces the largest scales of the flow, in particular in the vertical direction (see Arya (1975)) and thus increasing turbulence anisotropy (Armenio and Sarkar (2002)). Stable stratification also causes an increase of viscous dissipation (see Garg

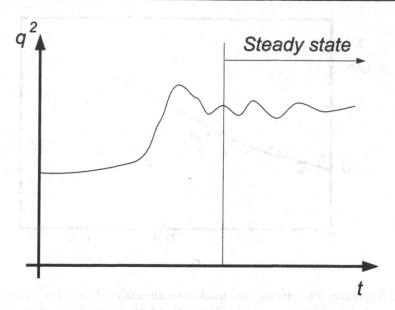

Figure 12. Typical time-behavior of the turbulent kinetic energy q^2 (also denoted with K in the present Chapter), during a DNS simulation. Reproduced with permission from Piomelli and Chasnov (1996).

(1996); Armenio and Sarkar (2002)) and, since the Kolmogorov length-scale is related to the dissipation rate according to:

$$\eta_k = (\frac{\nu^3}{\epsilon})^{1/4},$$

it follows that the dissipation length-scale tends to decrease. From this aspect, a grid resolution designed for simulating a neutrally stratified flow, may be not fine enough for the study of a stably stratified flow.

Achievements of DNS: DNS of the stably stratified homogeneous turbulence have been carried out by several authors (among them by Gerz et al. (1989); Herring and Metais (1989); Holt et al. (1992); Jacobitz et al. (1997) characterized by the presence of a constant vertical shear and a linear, vertical density gradient. The effect of horizontal shear in a stably stratified homogeneous flow has been investigated by Jacobitz and Sarkar (1998). Such flow is typically unsteady and time-evolving. Turbulent kinetic energy grows in time in subcritical conditions, when stratification is not strong enough to suppress turbulence; conversely, a decay of TKE in time is observed for strong stratification. Detailed discussion of physics of turbulent stratified flow analyzed by the use of direct simulation is given in Chapter 2 and Chapter 3 the present lecture notes. Herein we only briefly discuss some relevant achievements. A main result of Gerz et al. (1989) is that in presence of strong stratification, internal gravity waves build up superposed by

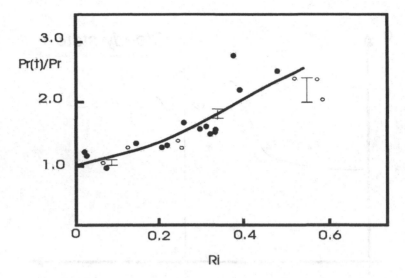

Figure 13. Turbulent Prandtl number made non-dimensional with the molecular value as a function of the Richardson number.Reproduced with permission from Gerz et al., (1989).

light fluid where counter-gradient buoyancy fluxes (CGBF) appear. Such CGBF have been shown to increase in intensity with the Prandtl number. Another result worth mentioning is that the turbulent Prandtl number increases with stratification (Fig. 13) thus confirming the experimental findings of Webster (1964) and the validity of the model of Launder (1975). The increase of the turbulent Prantdl number with overall stratification, means that under such conditions, momentum is much more likely to be transported by turbulence than density. The Reynolds analogy (that assumes a turbulent Prantdl number roughly equal to unity) is hence valid for passive scalars, but it does not hold for stably stratified flows. DNSs of particle dispersion in a homogeneous stably stratified flow have been carried out by Kimura and Herring (1996). The authors have found that vertical displacement of single particles (Fig. 14) and separation distance between pairs of particles are strongly inhibited by stable stratification. Conversely, dispersion in the horizontal planes appeared to be very weakly affected by stratification. This behavior confirms the increase of anisotropy associated to the increased stratification, and also observed in previous studies (see for instance Gerz et al. (1989); Herring and Metais (1989); Holt et al. (1992)). The analysis of homogeneous stably stratified flows has also allowed to identify the critical value of the gradient Richardson number Ri_g (defined in Chapter 3 of the present lecture notes) for the stability of the flow and, the flow parameters that affect $Ri_{g,crit}$ (Jacobitz et al. (1997); Jacobitz and Sarkar (1998)).

Shear driven flows of relevance in geophysical applications have been investigated by DNS. Coleman et al. (1990, 1992) have respectively studied the surface Ekman layer in

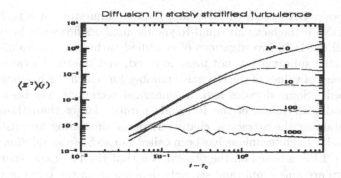

Figure 14. Vertical dispersion of massless particle in a homogeneous, stably stratified flow. Reproduced with permission from Kimura and Herring, (1996).

a neutral flow and in a stably stratified flow. This flow is governed by the rotational term appearing in the governing equations due to the rotation of the frame of reference. Such flow is considered to be the archetype of the planetary boundary layer. A main conclusion of Coleman et al. (1990) is that the horizontal component of vorticity cannot be neglected, since it is responsible of the production of a secondary Reynolds stress $\overline{u_1 u_2}$. The authors concluded that, due to the substantial influence played by the horizontal component of vorticity in turbulence activity, the $f - plane$ approximation cannot be adopted in the numerical simulation of the Ekman layer. In case of stable stratification, Coleman et al. (1992) have shown that the turbulent structure of the Ekman layer is modified by stratification. The modifications appear more evident in the deep core region, away from the free-surface, whereas, the free-surface region, characterized by the presence of strong shear stress is less affected by stratification. This has been attributed to the fact that buoyancy destruction primarily affects the large scales of the flow that are located in the core region. The penetration length of the Ekman layer has been found reduced by stratification. Finally stable stratification increases the angle of the shear stress whereas its magnitude appears to be unaffected.

DNS of stably stratified flows have also been conducted by several authors for the study of other important turbulence flows, as, for instance, the shear-layer. Among them, we cite the relevant investigations of Smyth and Moum (2000); Staquet (2000).

2.2 Modeling turbulence using large-eddy simulation

As outlined above, DNS is an effective tool for the simulation of a turbulent field. The drawback of DNS is the fact that small-Reynolds numbers flows can be investigated. For example, in all the DNS investigations of stratified turbulence above discussed, the presence of an inertial subrange has not been detected, and the results cannot be easily extended to the range of values of the Reynolds number for which the Kolmogorov theory has been developed. Some decades ago, a numerical technique has been developed, that is able to deal with values of the Reynolds number larger than those typical of a DNS, still maintaining the attractive characteristic of simulating an unsteady three-dimensional flow field. Such technique has been called *Large-Eddy simulation* (LES). The main philosophy of LES is based on the observation that the large and energy-carrying scales of turbulence are anisotropic and strongly dependent on the boundary conditions of the flow field (see Fig. 6), whereas the small and dissipative scales of turbulence exhibit characteristics of universality. These small scales also tend to be more and more isotropic and independent on the boundary conditions, *i.e.* on the way turbulence is generated in the flow field. Based on these considerations, the underlying idea was to solve directly the large scales of the flow by means of an unsteady and three-dimensional NS solver and to parameterize the large wavenumber dissipative scales by means of a subgrid-scale (SGS) model. From the physical point of view the process appears to be very clear. The large scales (small wavenumbers) where turbulence is generated are solved directly together with the left part of the inertial subrange, where the inviscid and non-linear transfer of energy down to the small scales takes place. The remaining of the spectrum is not directly solved rather a model is considered that is responsible for draining and dissipating the energy coming down from the large scales. This physical operation is mathematically performed applying a low-pass filter to the governing equations. The filtering operation is defined as:

$$\overline{f}(x_i) = \int_D f(x_i)\, G(x_i, x_{i,1})\, dx_{i,1}, \tag{2.8}$$

where the symbol $\overline{(.)}$ denotes the filtering operation, D is the spatial domain, $f(x_i)$ is the function to be filtered and $G(x_i, x_{i,1})$ is the filter function that determines the size and the structure of the small scales. The filter functions most commonly in use are the:
- sharp-Fourier cutoff filter, defined in the wavenumber space;
- Gaussian filter, defined in the real space;
- tophat filter, defined in the real space.

The sharp-Fourier cutoff is defined as:

$$\hat{G} = \begin{cases} 1 & if\ k \leq \pi/\Delta \\ 0 & \text{otherwise} \end{cases}$$

where the symbol $(\hat{.})$ denotes the Fourier coefficient of a quantity.

The Gaussian filter is defined in the real space as:

$$G(x) = \sqrt{\frac{6}{\pi\Delta^2}} e^{-\frac{6x^2}{\Delta^2}},$$

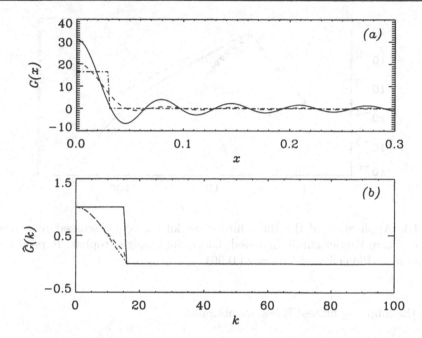

Figure 15. a) Filter functions in the real space: black line, sharp Fourier cutoff; gray line, Gaussian; dot line, tophat.
Filter functions in the wavenumber space: black line, sharp Fourier cutoff;gray line, Gaussian; dot line, tophat. Reproduced with permission from Piomelli and Chasnov (1996)

whereas the tophat filter has the following form:

$$G(x) = \begin{cases} 1/\Delta & if |x| \leq \Delta/2 \\ 0 & \text{otherwise} \end{cases}$$

where Δ is the filter width. The shape of the filter functions in the real space and in the Fourier space are shown in Figs. 15.

The difference between the filters can be detected by filtering a test function that has a well defined spectrum (the solid line of Fig 16) in the wavenumber space. Filtering is performed using the above defined filter functions and the same filter amplitude. Figure 16 shows that the sharp cutoff filter gives a sharp truncation of the spectrum, whereas the filters defined in the real space tend to affect a wide range of scales at wavenumbers smaller than that of truncation. If the filtering operation is applied to the Navier-Stokes

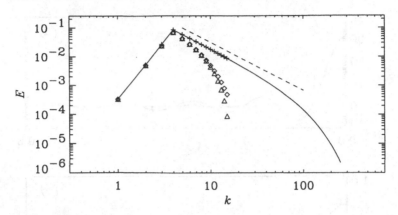

Figure 16. Application of the three filter functions herein discussed to a test function: cross, sharp Fourier cutoff; diamond, Gaussian; triangle, tophat. Reproduced with permission from Piomelli and Chasnov (1996).

equation the following filtered NSEs are obtained:

$$\frac{\partial \overline{u_i}}{\partial x_i} = 0, \tag{2.9}$$

$$\frac{\partial \overline{u_i}}{\partial t} + \frac{\partial \overline{u_j}\,\overline{u_i}}{\partial x_j} = -\frac{1}{\rho_0}\frac{\partial \overline{p}}{\partial x_i} + \nu \frac{\partial^2 \overline{u_i}}{\partial x_j \partial x_j} - \frac{\overline{\rho_p}}{\rho_0}g\delta_{i3} - \frac{\partial \tau_{i,j,sgs}}{\partial x_j} \tag{2.10}$$

$$\frac{\partial \overline{\rho_p}}{\partial t} + \frac{\partial \overline{u_j \rho_p}}{\partial x_j} = \frac{\nu}{Pr}\frac{\partial^2 \overline{\rho_p}}{\partial x_j \partial x_j} - \frac{\partial \lambda_{j,sgs}}{\partial x_j} \tag{2.11}$$

The equations are formally similar to the Reynolds averaged NSE, although their physical meaning is profoundly different. In Eqs 2.10,2.11 the terms that appear at the right-hand-side as a result of the filtering operation, respectively are the subgrid-scale stress tensor and the subgrid-scale density fluxes. The SGS stresses are:

$$\tau_{i,j,sgs} = \overline{u_i u_j} - \overline{u_i}\,\overline{u_j},$$

whereas, the SGS density fluxes are:

$$\lambda_j = \overline{\rho_p u_j} - \overline{\rho_p}\,\overline{u_j}.$$

Obviously, if one is working with the transport equation of the temperature and/or concentration (such as salinity) equivalent SGS fluxes can be derived.

It is worth paying attention to the differences between the RANS and the filtered LES equations:

- the RANS equations govern the transport of the *mean* field and the Reynolds stress tensor accounts for the contribution of *all* scales of turbulence to the evolution of the mean field;

- the LES equations govern the evolution of the three-dimensional and unsteady filtered field, and the SGS stress tensor accounts for the contribution of the *unresolved*
  scales to the resolved ones. Since the SGS stress contains the contribution of the
  small scales, the spectral analysis discussed in a previous section, suggests that they
  amount to a small percentage of the total Reynolds stresses. Similar arguments
  hold for the density fluxes.

In order to go deep in the comprehension of the differences between the Reynolds-average
approach and the filtered one, it is useful to write the relationships between mean and
filtered variables.

In LES, the total field (velocity, pressure and perturbation density) is decomposed
into two contributions, namely a filtered one and a SGS contribution:

$$f(x_i, t) = \overline{f}(x_i, t) + f_{sgs}.$$

The filtered part can be further decomposed in a RANS-sense, to give

$$\overline{f}(x_i, t) = F(x_i) + f''(x_i, t)$$

where the capital letter indicates a Reynolds-averaged value. In other words the filtered
field can be decomposed into a mean value and a fluctuating value that represent the
resolved fluctuating field. Figure 17 clearly shows the differences between a total turbulent field (obtained in this case by the use of a DNS), a filtered field (obtained applying
the sharp Fourier cutoff) and the mean value: it is possible to see that the filtered field
contains most of the fluctuation and that the difference between the total and the filtered
quantity is mostly confined within the small scales. Hereafter Reynolds averaging is
indicated with the symbol $< \, . \, >$, whereas an overline indicates the filtering operation.
It is also possible to find a relationship between the total Reynolds stresses and the SGS
quantities. In particular it can be easily shown that:

$$< u_i' u_j' > = < u_i'' u_j'' > + \tau_{ij,sgs},$$

and similarly for the density fluxes:

$$< u_j' \rho_p' > = < u_j'' \rho_p'' > + \lambda_{j,sgs}.$$

The SGS stresses and density fluxes need to be modeled. However, the hope that has
pushed most of the research in LES, is that, based on the local isotropy of the small
scales of turbulence and on their own universality, a relatively inexpensive and simple
model can be able to work satisfactory.

It is instructive to look at the equations of evolution of the *filtered* turbulent kinetic
energy $\overline{q}^2 = \overline{u}_i \, \overline{u}_j$, and of the SGS counterpart $q_{sgs}^2 = \tau_{sgs,ii}$:

$$
\frac{\partial \overline{q}^2}{\partial t} + \frac{\partial \overline{q}^2 \overline{u}_j}{\partial x_j} = -2 \frac{\partial \overline{p} \, \overline{u}_j}{\partial x_j} + \nu \frac{\partial^2 \overline{q}^2}{\partial x_j \partial x_j} - 2 \frac{\partial}{\partial x_j} (\tau_{ij} \overline{u}_i) - \tag{2.12}
$$

$$
-2\nu \frac{\partial \overline{u}_i}{\partial x_j} \frac{\partial \overline{u}_i}{\partial x_j} + 2\tau_{sgs,ij} \overline{S}_{ij} - 2 \frac{g}{\rho_0} \overline{\rho_p} \, \overline{u}_j \delta_{j3}
$$

The terms of Eq. 2.12 have the following meaning:

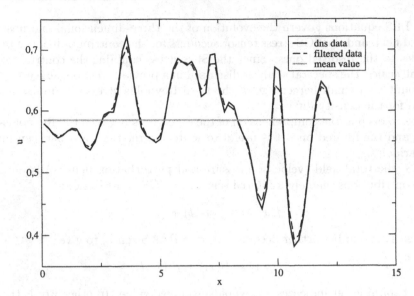

Figure 17. Space distribution of the streamwise component of the velocity field at a given distance from the wall in the simulation of a channel flow: solid line, DNS data (128 grid points); dashed line, filtered data, obtained filtering the DSN data over a $2\Delta x$ filter width; gray bold line, Reynolds averaged value.

- the second term of the left-hand-side is advection of $\overline{q}^2$ by the filtered velocity field;
- the first two terms of the RHS are respectively diffusion of $\overline{q}^2$ by filtered pressure and by viscosity;
- the third term of the RHS id SGS diffusion;
- the fourth term of RHS is viscous dissipation of $\overline{q}^2$;
- the fifth term of RHS is dissipation of $\overline{q}^2$ to fed the SGS scales;
- the sixth term is a buoyancy term given by interaction between filtered density and vertical velocity. Once again, this term can supply or drain energy depending on the kind of stratification (respectively unstable or stable).

The transport equation of q_{sgs}^2 reads as:

$$\frac{\partial \overline{q}_{sgs}^2}{\partial t} + \frac{\partial \overline{q}_{sgs}^2 \overline{u}_j}{\partial x_j} = -\frac{\partial}{\partial x_j}(\overline{u_i u_i u_j} - \overline{u_i u_i}\ \overline{u_j}) - 2\frac{\partial}{\partial x_j}(\overline{pu_j} - \overline{p}\ \overline{u_j}) + \nu\frac{\partial^2 \overline{q}_{sgs}^2}{\partial x_j \partial x_j} + \quad (2.13)$$

$$+2\frac{\partial}{\partial x_j}(\tau_{sgs,ij}\overline{u}_i) - 2\nu(\overline{\frac{\partial u_i}{\partial x_j}\frac{\partial u_i}{\partial x_j}} - \frac{\partial \overline{u}_i}{\partial x_j}\frac{\partial \overline{u}_i}{\partial x_j}) - 2\tau_{sgs,ij}\overline{S}_{ij} - 2\frac{g}{\rho_0}\lambda_3$$

In Eq. 2.13:
- the second term of the LHS is advection of q_{sgs}^2 by filtered field;
- the first term of RHS is turbulent transport of q_{sgs}^2 by triple-correlation;
- the second and third terms of RHS are respectively pressure and viscous diffusion of q_{sgs}^2;

- the fourth term of RHS is SGS diffusion;
- the fifth term is viscous dissipation of q_{sgs}^2;
- the sixth term is production of q_{sgs}^2 by interaction between resolved and unresolved scales.
- the seventh term is production or destruction (respectively in unstable or stable stratification) of q_{sgs}^2 by SGS buoyancy fluxes.

The equations given above show that the filtered scales interact with the unresolved ones in a control volume through several mechanisms. As discussed, advection and diffusion represent pure redistribution of energy. The $4th$ and the $5th$ term provide to the dissipation of $\overline{q}^2$. The former gives dissipation at the resolved level, whereas, the latter takes into account for the net energy exchange between the resolved and the unresolved scales. Although such term can be either positive or negative, on average it is negative, meaning that energy is drained by the small scales and the energy flows from the large to the small scales (forward-scatter). Positive values are associated to the presence of back-scatter, that means energy going up from the small to the large scales. This phenomenon has been observed to occur locally, although, on average the energy flows down toward the small scales. As regards the subgrid-scale kinetic energy, a similar discussion can be done. In this case $\tau_{sgs,ij}\overline{S}_{ij}$ acts as a source term, since the energy lost by the large resolved scales feds the unresolved small ones.

Similar equations can be derived for the perturbation density.

models for the sgs stresses and density fluxes   Once the filtered Navier-Stokes equations are written and the mechanism of transport of energy from the resolved scales to the SGS ones understood, we need a closure for the SGS terms. A significant help in the development of the SGS models comes from the general theory of the equilibrium range and, in particular from the observation that the dissipation rate of turbulent kinetic energy can be inferred by the knowledge of the energy content at the large scales. This implies that, if one captures the large scales accurately, the model has just to be able to drain and to dissipate energy coming down from those scales, and it does not need to be very sophisticated (and CPU-time consuming). In the past, this point has lend justification to a very crude approach of using a dissipative numerical scheme, in conjunction with a not resolved DNS, leaving to the numerical dissipation the role of dissipating the energy coming from the large scales. Unfortunately, the use of a dissipative numerical scheme affects the energy spectrum well above the cut-off, and it causes a wrong decay of turbulent kinetic energy in the inertial range. Figure 18 shows the results from numerical experiments carried out by Mittal and Moin (1997), aimed at exploring the conservation properties of high-order upwind schemes used for the discretization of spatial derivatives. In particular, in Fig. 18 it can be observed that upwinding always causes a wrong decay of TKE, no matter of the order of discretization, whereas $2nd$-order centered schemes give a correct slope of the energy spectrum up to the cutoff. In other words, the use of upwinding roughly corresponds to simulate a much smaller Reynolds-number flow, characterized by a much larger value of the spatial Kolmogorov scale. In general, successful LES have been carried out using conservative (not dissipative) numerical schemes in conjunction with SGG models.

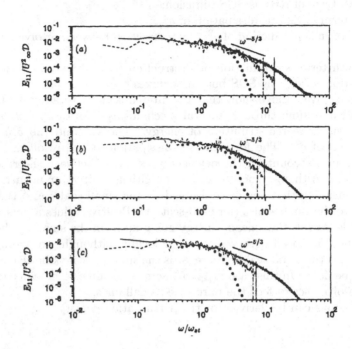

Figure 18. Spanwise spectrum of the streamwise energy for the LES of the turbulent flow over a circular cylinder. Solid line, experimental data; dashed line, data form LES with 2nd-order centered scheme; dotted line, data from LES with 5th-order upwind scheme. The figures refer to three different longitudinal sections in the wake region. Reproduced with permission from Mittal and Moin (1997)

Over the past decades a large number of SGS models have been developed and used for the simulation of turbulent flows. Citing all these models is beyond the scopes of the present lectures, rather we try to focus on few models that have been proved to work satisfactorily for the simulation of neutral and stratified flows. To start with, models have been defined in spectral space and in the physical space. Since the latter are most commonly used, we refer to the latter.

Most models have been developed using the eddy-viscosity concept, similarly to RANS approach. In particular, the SGS stresses can be modeled as:

$$\tau_{sgs,ij} - \frac{\delta_{ij}}{3}\tau_{sgs,kk} = -2\nu_T \overline{S}_{ij} \qquad (2.14)$$

that makes the SGS stresses proportional to the resolved strain-rate tensor

$$\overline{S}_{ij} = \frac{1}{2}\left(\frac{\partial \overline{u}_i}{\partial x_j} + \frac{\partial \overline{u}_j}{\partial x_i}\right)$$

Once Eq. 2.14 is accepted, the problem is moved to the evaluation of the *sub-grid scale* eddy viscosity. A simple way to obtain ν_T is to use an algebraic model, thus avoiding to increase the computational cost of an already expensive simulation. The use of a simple model is justified by several arguments already discussed and herein clearly evidenced:
- the small scales tend to be more and more homogeneous and isotropic;
- since most scales are resolved directly, and only the small not energetic scales are modeled, modeling errors are expected to be less significant in the overall accuracy of the simulation.

Eddy-viscosity models have been developed based on dimensional arguments, as done within the RANS framework in the past. The eddy viscosity can be considered as the product of a length-scale and a velocity-scale. In LES, the typical length-scale is given by the filter width, that gives the size of the smallest resolved scales, whereas a velocity scale can be taken as $\sqrt{q^2_{sgs}}$.

$$\nu_t \sim \Delta \sqrt{q^2_{sgs}}$$

Different approaches can be followed for deriving the SGS kinetic energy. The simplest is to take advantage of the assumption of local equilibrium (production=dissipation) to arrive to an algebraic expression. As discussed in a previous chapter, the equilibrium assumption is based on the observation that the small scales evolve (and thus adjust to local equilibrium) much more rapidly than the large ones. Under this assumption, most terms of the transport equation 2.13 drop out and the following simplified form holds:

$$\epsilon_{sgs} = -\tau_{sgs,ij}\overline{S}_{ij}$$

In simple flows, this hypothesis has been shown to work satisfactorily (Bardina et al. (1985)), although, in more complex flows, characterized by departure from equilibrium, local back-scatter etc., the assumption of local equilibrium may be not appropriate and more sophisticated models are required.

The pioneering eddy-viscosity model was developed by Smagorinsky (1963), who has used the mixing length concept to relate the SGS eddy viscosity to the resolved strain-rate tensor as follows:

$$\nu_T = (C_s\Delta)^2|\overline{S}|$$

where $C_s\Delta$ is a length-scale proportional to the filter width and $|\overline{S}| = \sqrt{2\overline{S}_{ij}\overline{S}_{ij}}$ is the contraction of the filtered strain-rate tensor. The above formulation was originally developed for *quasi* two-dimensional flows, in that only the horizontal component of the strain-rate tensor was used. In a successive work, Lilly (1967) found out that the original Smagorinsky's formulation holds in three-dimensional flows and determined the constant C_s, based on the assumption that the filter width lies within the inertial subrange. In particular, using the Kolmogorov law of the inertial subrange, and assuming that the all the dissipation is contained in the unresolved scales, the author has found:

$$C_s = 0.18$$

This parameterization has worked reasonably well in unbounded, neutrally stratified turbulence. It does not account for the effects of stratification and rotation on the small scales. In particular, Deardoff (1973) has found the Smagorinsky model to be inaccurate when applied to the study of the stably stratified planetary boundary layer. To summarize, the Smagorinsky model used in conjunction with the Lilly's findings, can give reasonable results in very simple flow configurations. It is not suited for geophysical flows where effects of stratification rotation and local intermittency are likely to take place.

Another problem of the Smagorinsky model is its inapplicability to the study of wall-bounded turbulence. Indeed, the model gives a very large value of eddy viscosity at the wall, where it is expected to be zero. In other words, the Smagorinsky model does not take into account the reduction of the turbulent length-scale when moving toward the wall. The problem has been overcome by artificially reducing the near-wall length scale by means of a damping function. In this case the eddy viscosity is:

$$\nu_T = [C_s\Delta(1 - exp(-z^+/25))]^2|\overline{S}|$$

where $z^+ = z/z^*$ is the distance from the wall made dimensionless with the wall unit $z^* = u_\tau/\nu$ with $u_\tau = \sqrt{\tau_w/\rho_0}$ the friction velocity and τ_w the wall stress. Although the wall adjustment has allowed to obtain reasonable results in wall bounded turbulence, the performances of the model have appeared to be sensitive to the proper tuning of th constant. In the application of the Smagorinsky model, the filter width Δ is chosen to be equal to the grid size, and thus an implicit filtering is performed just solving the filtered Navier-Stokes equation using a grid spacing much larger that the Kolmogorov microscale.

Some drawbacks of the Smagorinsky model have been overcome by introducing the *dynamic* models. The underlying idea of such models consists in a *dynamic* evaluation of the constant, based on the physical characteristics of the turbulent field. In particular the evaluation of the constant is based on the energy content of the smallest resolved scales. The dynamic model has been first developed by Germano et al. (1991) and successively improved by other authors. Within the dynamic framework, two filters are used, namely a grid filter (denoted by a bar) and a test filter (denoted by a caret); the width of the test filter $\widehat{\Delta}$ is larger than that of the grid filter Δ.

As already discussed the application of a filter Δ to the NSE produces the SGS stress $\tau_{sgs,ij}$. Now, if we apply a test filter (see Fig. 19) to the filtered NSE we obtain the following subtest-scale stresses $T_{sts,ij} = \widehat{\overline{u_i u_j}} - \widehat{\overline{u}}_i\widehat{\overline{u}}_j$. Germano (1992) has shown that the following identity holds:

$$L_{ij} = \widehat{\overline{u}_i\overline{u}_j} - \widehat{\overline{u}}_i\,\widehat{\overline{u}}_i = T_{sts,ij} - \widehat{\tau}_{sgs,ij}$$

that gives a relationship between the resolved turbulent stress L_{ij} the subtest stress $T_{sts,ij}$ and the SGS stress $\tau_{sgs,ij}$. Note that the quantity L_{ij} can be calculated explicitly. Now a choice must be done for the parameterization of the stresses T and τ. Assuming the existence of an inertial sub-range in the energy spectrum, it is possible to use the same parameterization at the test and grid levels. For example, the use of an eddy viscosity

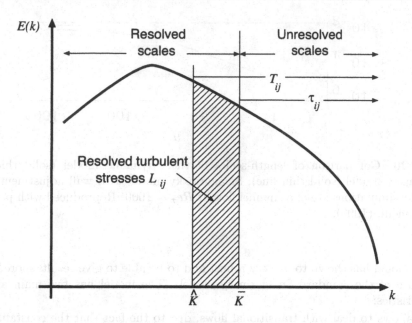

Figure 19. Schematic of test filtering and grid filtering in the energy spectrum. Reproduced with permission from Piomelli and Chasnov (1996).

model gives:

$$\tau_{sgs,ij} - \frac{\delta_{ij}}{3}\tau_{sgs,kk} = -2C\Delta^2|\overline{S}|\overline{S}_{ij} = -2\,C\,m_{ij}; \qquad (2.15)$$

$$T_{sts,ij} - \frac{\delta_{ij}}{3}T_{sts,kk} = -2C\widehat{\Delta}^2|\widehat{\overline{S}}|\widehat{\overline{S}}_{ij} = -2\,C\,n_{ij}. \qquad (2.16)$$

Substitution of the expressions above in the Germano's identity gives the following set of five independent equations:

$$L_{ij}^a = L_{ij} - \frac{\delta_{ij}}{3}L_{kk} = -2Cm_{ij} - 2\widehat{Cn_{ij}} \qquad (2.17)$$

In order to obtain a single coefficient, Lilly (1992) proposed to take out the coefficient C from the filtering operation and to minimize the sum of the squares of the residual of Eq. 2.17 by contracting the terms of the equation with $m_{ij} - \widehat{n}_{ij}$. This operation gives the following algebraic expression:

$$C(x_i, t) = -\frac{1}{2}\frac{L_{ij}^a(m_{ij} - \widehat{n}_{ij})}{(m_{ij} - \overline{n}_{ij})^2} \qquad (2.18)$$

The coefficient is thus calculated dynamically during the computation and calibration is no longer required.

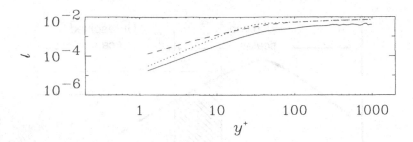

Figure 20. Comparison of length-scales of the: dynamic model (solid thick line); Smagorinsky model (solid thin line); Smagorinsky model with wall adjustment (dotted line). The simulations are of a channel flow at $Re_\tau \sim 1000$. Reproduced with permission from Piomelli, (1993).

Such model has shown to be very robust and to be able to give results more accurate than those of the standard Smagorinsky model. The model has two main attractive characteristics:

- it allows to deal with transitional flows, due to the fact that the constant goes to zero in the laminar regions (characterized by absence of small scale fluctuations);

- in wall bounded flows, the constant goes to zero in the near-wall region without the need to use any damping function;

Applications of the dynamic model to wall-bounded turbulence (Piomelli (1993) have shown results much more accurate than that obtained with the Smagorinsky model with the van-Driest function, basically because the dynamic estimation of the constant give a more accurate prediction of the near-wall length-scale compared to the empirical damping function (Fig. 20).

A drawback of the Lilly's estimation of the constant C stands in the fact that the constant is arbitrarily extracted from a filtering operation. This operation is mathematically not consistent. However, consistence is recovered if the constant is averaged over the directions of homogeneity. In non-homogeneous flows, special localization techniques can be used (see for example the Lagrangian model of Meneveau et al. (1996)), to make the evaluation of the constant mathematically consistent.

Finally, since the dynamic model requires at least an explicit operation of filtering, problems can arise when the grid is not regular along the directions of explicit filtering. This problem has been easily overcome in the study of the channel flow or of the boundary layer over a flat wall. Indeed, in these cases, filtering has been applied over the directions of homogeneity where the grid is in general uniform. For complex geometry, a re-visitation of the filtered NSE, in curvilinear coordinates has been performed by Jordan (1999). The author has shown that is possible to perform filtering in the computational space (that is a rectangular box with uniform grid spacing in all directions) provided that the contravariant fluxes (instead of the velocity components) and the metric terms are filtered. Such approach requires the construction of contravariant quantities, their filtering in the computational space and to go back to the physical velocity components.

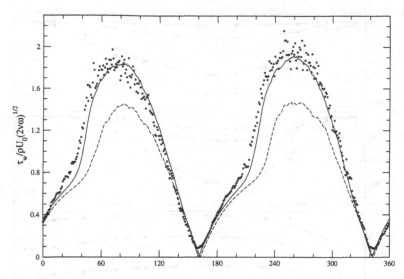

Figure 21. Phase-averaged values of the wall stress during a cycle of oscillation in the Stokes boundary layer at $Re_\delta = 1790$. The results of a coarse grid simulation (dashed line) and of a fine grid simulation (black line) are compared with those of the experiments of Jensen et al., 1989. Reproduced with permission from Salon, (2004).

Nonetheless, it allows to use a mathematical consistent dynamic model for the analysis of complex geometry. The dynamic model has been used in a very wide variety of problems and a detailed description of such simulation is beyond the scope of the present notes. A very recent application of the dynamic model to the study of complex flows has been done by Salon (2004). The authors have simulated the oscillatory boundary layer over a flat bottom using the dynamic mixed model formulated in Armenio and Piomelli (2000). The model contains a scale-similar part (Bardina et al. (1980)) an eddy viscosity part and is treated dynamically as shown above. The complexity of the flow stands in the fact that, at the Reynolds number of the simulation ($Re_\delta = 1790$, where δ is the nominal thickness of the Stoke's boundary layer) both sharp transition to turbulence and rapid decay to a transitional state occur along a cycle. Triggering to turbulence appears during the early stage of the acceleration phase; turbulence is sustained over the remaining of the acceleration phase of the cycle up to the late deceleration phase where it decays.

Figure 21 shows that a dynamic SGS model, used with a fine grid, is able to give a satisfactory fitting of the time evolution of the wall stress measured in the experiments. The sharp transition to turbulence, occurring where a sharp increase of the wall stress is observed, is also well reproduced. Figure 22 shows that the SGS model is not active during the late deceleration and early acceleration phases, where turbulence is not active, whereas it switches on and supplies SGS dissipation and eddy viscosity in the remaining of the cycle where turbulence is active.

The SGS closures described above have been extended to the case of stratified flows.

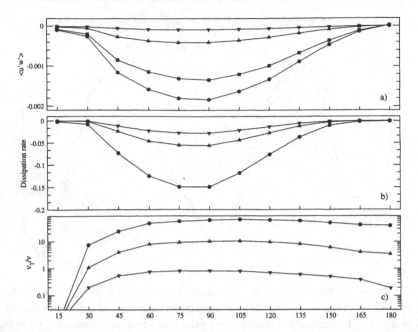

Figure 22. Mean values throughout the cycle: a): Reynolds stress at $z_d \approx 0.5\delta_S$: total $\langle u'w' \rangle$ ($\bullet$), $\langle u'w' \rangle - \langle u' \rangle \langle w' \rangle$ ($\blacksquare$), $\langle u'w' \rangle_{ss}$ ($\blacktriangle$), $\langle u'w' \rangle_{ev}$ ($\blacktriangledown$); b) dimensionless dissipation at $z_d \approx 0.15\delta_S$: total ($\bullet$), ss component ($\blacktriangle$), ev component ($\blacktriangledown$); c) normalized turbulent viscosity at $z_d \approx 2\delta_S$: total ($\bullet$), ss component ($\blacktriangle$), ev component ($\blacktriangledown$). Reproduced with permission from Salon, (2004).

The simplest model for treating the SGS density fluxes is the following:

$$\lambda_j = -k_T \frac{\partial \rho_p}{\partial x_j} \qquad (2.19)$$

where the SGS turbulent diffusivity is expressed as $k_T = C_\rho \Delta^2 |\overline{S}|$. Eidson (1985) has expressed the constant C_ρ as a function of C as

$$C_\rho = \frac{C}{Pr_T}$$

where Pr_T is the SGS Prandtl number. The SGS model has two free parameters, namely C and Pr_T, that need to be calibrated depending on the physical characteristics of the flow field. A similar model has been used by Kaltenbach et al. (1994) for the simulation of the homogeneous stably stratified flow of Gerz et al. (1989) at a larger value of the Reynolds number. In their simulation, the Smagorinsky constant has been set equal to $C = 0.17$ and two different values of the SGS Prandtl number have been used, namely $Pr_T = 0.5$ and $Pr_T = 1$ with the aim to check the sensitivity of the results on this parameter. It is noteworthy that the resolved strain rate tensor used for computing the SGS terms did not include the mean shear that drives the motion. As regards the performances of the model the authors have found that the results of LES were in good agreement with those of DNS in case of weak stratification, namely for values of the gradient Richardson number $Ri_g = N^2/S^2$ (with N the Brunt-Vaisala frequency defined in Chapter 2 of the present lecture notes and S the mean shear) smaller that the critical value of 0.25; on the other hand for supercritical values of Ri_g a progressive departure of the LES data from the reference DNS ones has been observed. For instance in decaying turbulence occurring for $Ri_g > 0.25$ the counter-gradient buoyancy fluxes observed in the DNS simulation have not been reproduced by the LES model employed. The sensitivity of the results on the variation of the SGS Prandtl number has been found to be more and more pronounced for supercritical conditions.

The underlying idea of the dynamic model of Germano et al. (1991), namely the dynamic evaluation of the constant during the computation, has been successfully extended to the case of stratified flows. In particular, the application of the test-filter produces the following subtest scale density fluxes:

$$\Lambda_i = \widehat{\overline{\rho_p u_i}} - \widehat{\overline{\rho_p}} \, \widehat{\overline{u_i}}$$

and the the following Germano-like identity holds:

$$F_i = \widehat{\overline{\rho' \, \overline{u_i}}} - \widehat{\overline{\rho'}} \, \widehat{\overline{u_i'}} = \Lambda_i - \widehat{\lambda_i}.$$

The use of an eddy viscosity model for the subtest as well as subgrid density fluxes, leads to:

$$\lambda_i = -2 C_\rho \Delta^2 |\overline{S}| \frac{\partial \overline{\rho_p}}{\partial x_i} = -2 \, C_\rho \, p_i; \tag{2.20}$$

$$\Lambda_i = -2 C_\rho \widehat{\Delta}^2 |\widehat{\overline{S}}| \frac{\partial \widehat{\overline{\rho_p}}}{\partial x_i} = -2 \, C \, q_i. \tag{2.21}$$

Applying the technique of Lilly (1992) we obtain the following expression:

$$C_\rho = -\frac{1}{2} \frac{F_i(q_i - \widehat{p_i})}{(q_i - \widehat{p_i})^2} \tag{2.22}$$

A detailed description of the dynamic eddy diffusivity model is given in Cabot and Moin (1993). *A priori* tests using DNS databases of homogeneous turbulence and wall bounded turbulence for the case of passive scalar with different Prandtl numbers, have shown that the dynamic model is able to give a reasonable prediction of SGS Pr_T. In particular, Pr_T has been shown to be weakly dependent on the molecular Prandtl number, and nearly equal to 0.5 in homogeneous turbulence and in the outer region of a boundary layer. Finally, comparison of LES results with DNS data have shown that the dynamic eddy diffusivity model, implemented with the Lilly's procedure gives accurate results of the first and second order statistics of the turbulent field.

The dynamic eddy diffusivity model has been successfully employed for the simulation of stably stratified flows. The LES investigation of the stably stratified turbulent channel flow by Armenio and Sarkar (2002) has shown the ability of the dynamic model in

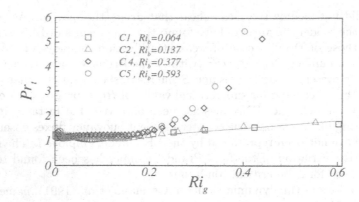

Figure 23. The influence of the gradient Richardson number on the total turbulent Prandtl number for different levels of stratification. Reproduced with permission from Armenio and Sarkar, (2002).

simulating stably stratified flows even in cases of strong stratification. Indeed, the model is able to give *automatically* the increase of the total and subgrid Prandtl numbers with stratification. Fig. 23 shows the total Prandtl number:

$$Pr_{tot} = \frac{< u'w' > \frac{d<u>}{dz}}{< \rho'w' > \frac{d<\rho>}{dz}} .$$

as a function of the gradient Richardson number Ri_g for different simulations, from neutral to strongly stratified flow. For a given overall level of stratification, Ri_g increases going from the wall toward the centerline of the channel.

Figure 23 shows a linear and weak increase of the total Prandtl number with Ri_g for weak stratification, whereas a rapid, super-linear increase of Pr_T with Ri_g occurs for strong overall stratification. The SGS Prandtl number behaves in a similar way (Fig. 24). The two Figures show a very interesting feature: for subcritical conditions $Ri_g < 0.25$ the two Prandtl numbers are nearly constant, whereas for supercritical conditions ($Ri_g > 0.25$) a rapid and super-linear behavior of Pr_{tot} and Pr_T is observed. This finding may be helpful in explaining why in the simulation of Kaltenbach et al. (1994) the results were somewhat insensitive to the value of Pr_T in subcritical conditions and very strongly affected by the choice of Pr_T in supercritical conditions. A similar behavior has been observed in a successive study where the direction of the main shear was orthogonal to that of the density gradient (see Armenio and Sarkar (2004)).

The subgrid-scale models discussed above, (in particular the dynamic one) have been successfully applied to the study of idealized, small-scale problems. In particular, when analyzing wall-bounded turbulent flows, the Reynolds number of the simulation has always been set small enough to resolve most of the energy spectrum. It is well established

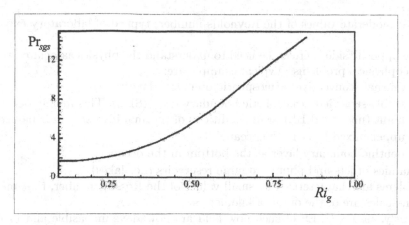

Figure 24. The influence of the gradient Richardson number on the SGS Prandtl number for moderate stable stratification. Reproduced with permission from Armenio and Sarkar, (2002).

that a turbulent boundary layer is characterized by the presence of two different regions. An outer (far-field) region, dominated by the presence of large-scale eddies that is independent on the fluid viscosity, and an inner (near-wall) zone where viscosity is important. The near-wall region is characterized by the presence of quasi-streamwise vortices that directly affect the evolution and the magnitude of the wall stress. The dimension of such vortices, when scaled with the inner-layer length-scale $z^* = \nu/u_\tau$ is nearly independent on the Reynolds number of the flow field. It can be easily shown that z^* decreases with Re, hence, the dimension of the near wall structures decreases with Re. This poses serious limitations to the application of LES to wall-bounded high Reynolds number flows, unless a wall model is used for the parametrization of the inner layer. Recently, the definition of *resolved* LES has been given to simulations that *directly* resolve the boundary layer up to the wall. Since the near wall structures are about 1000 wall units in the streamwise direction and about 100 wall units in the spanwise direction, a criterion for an accurate LES of wall turbulence is that the grid spacing made dimensionless with z^* are $\Delta x^* < 70$ and $\Delta z^* < 20$ respectively in the streamwise and spanwise direction. Moreover the first velocity point in the wall normal direction is required to be placed within the viscous sublayer (at $z/z^* = 1$) and about 8 points are needed within $z/z^* = 1$. Such stringent requirements, have allowed the application of *resolved* LES to flow field characterized by values of the Reynolds number of the order 20000 to 50000. To the best of our knowledge, the largest wall-bounded flow investigated by *resolved* LES is a channel flow at centerline Reynolds number $Re_c = 105000$ (see Kravchenko et al. (1996)). The authors have developed a very complicated numerical method that uses embedded grids in the near wall region aimed at giving enough resolution of the near wall structures, in conjunction with a dynamic eddy viscosity model. In stably stratified flows, the vertical length-scale is further reduced and it implies the need of more vertical resolution. As a consequence, *resolved* LES is able to solve wall-bounded turbulent flows at small-to-moderate values of the Reynolds number, typical of laboratory experiments.

On the opposite side there is the need to understand the physics associated to a broad class of geophysical problems. Typical example are:

- the diurnal (convective) atmospheric boundary layer;
- the stably stratified atmospheric boundary layer (SBL). This mainly occurs during the night (nocturnal BL) or in conditions of thermal inversion in urbanized areas.
- the upper mixed layer in the ocean;
- the benthic boundary layer at the bottom in the ocean;
- dynamics of jets and plumes in large reservoirs (i.e. lakes)

Such problems are characterized by small values of the Rossby number, that means that the length-scales are of the order of kilometers.

Obviously, *resolved* LES of such flow field are nowadays unfeasible and proper subgrid scale models have been developed for dealing with such problems. Due to its own importance, large efforts have been devoted to the study of the atmospheric boundary layer (ABL). The convective (unstably stratified) and neutral ABL have been successfully studied using LES by many authors, among them by Moeng (1984); Moeng and Sullivan (1994); Sullivan et al. (1994, 1996); Mironov et al. (2000). Comparatively, few LES of oceanic convection have been carried out. A key simulation of deep convection driven by surface buoyancy fluxes has been carried out by D.W.Denbo and Skyllingstad (1996). Moeng and co-workers used an eddy viscosity-diffusivity model composed of two parts for the parameterization of the turbulent stresses and of a single part for the SGS density fluxes:

$$\tau_{ij} = -2\nu_t \overline{S}_{ij} - 2\nu_T \langle S_{ij} \rangle,$$

$$\lambda_j = -k_t \frac{\partial \overline{\rho_p}}{\partial x_j}.$$

The SGS eddy viscosity has been expressed as:

$$\nu_t = C \, l_s q_{sgs}^{1/2}$$

where l_s is a mixing length , $q_{sgs}^{1/2}$ is a velocity scale dependent on the SGS turbulent kinetic energy q_{sgs} and C is a constant whose value was set equal to 0.1.

The SGS diffusivity was set dependent on the SGS eddy viscosity as:

$$k_t = \frac{\nu_t}{Pr_T}$$

with the SGS Prandtl number:

$$Pr_T = \frac{1}{1 + \frac{2\,l_s}{\Delta}}$$

The choice of the mixing length depends on the sign of stratification. For convective conditions, turbulence is enhanced by stratification and a proper choice is the filter width that is strictly related to the mesh size:

$$l_s = \Delta = [(9/4)\Delta_x \Delta_y \Delta_z]^{1/3},$$

where the coefficient 9/4 has been introducing due to the de-aliasing of the upper 1/3 of wavenumber in the horizontal direction where pseudo-spectral evaluation of the spatial derivatives is performed. In case of stable stratification, according to Deardoff (1980), the mixing length is modified as follows:

$$l_s = min(\Delta, 0.76 \, q_{sgs}^{1/2}/N)$$

thus introducing a length-scale $q_{sgs}^{1/2}/N$ dependent on the local level of stratification, the latter quantified by N. The second part of the model for the turbulent stresses has a RANS-like expression, and is used for ensuring a correct behavior at the wall:

$$\nu_T = \nu_T^* \frac{k_v z_1}{u_\tau \Phi_m(z_1)} |\langle S \rangle| \; for \; z \neq z_1 \tag{2.23}$$

$$\nu_T = \nu_T^* \; for \; z = z_1 \tag{2.24}$$

with z_1 the first grid point off the surface and Φ_m a function that depends on the stability of the system:

$$\frac{d\langle u \rangle}{dz} = \frac{u^* \Phi_m}{k_v z}$$

where k_v is the von Karman constant. The near surface mean eddy viscosity is:

$$\nu_T^* = \frac{u_\tau k_v z_1}{\Phi_m(z_1)} - \langle \nu_t \gamma \rangle - \frac{k z_1}{u_\tau \Phi_m(z_1)} (\langle u'w' \rangle_{z=z_1}^2 + \langle v'w' \rangle_{z=z_1}^2)^{1/2}$$

The isotropic factor γ takes into account for anisotropic effects near the surface and is used to shift the model from LES-type to a RANS-type in the surface region. For details the reader is referred to Saiki et al. (2000).

Unlike the Smagorinsky model, the SGS velocity scale is obtained solving a transport equation for the SGS turbulent kinetic energy q_{sgs}. Nieuwstadt and Brost (1986) have also developed a very popular SGS, eddy viscosity model for the analysis of the stratified atmospheric boundary layer. The model used the eddy viscosity-diffusivity approach and the velocity scale has been determined solving a transport equation of the SGS turbulent kinetic energy. Obviously the use of such an equation requires some assumptions for the modelization of transport terms of q_{sgs}. The performances of the LES models commonly in use for the analysis of the convective (and neutral) ABL have been evaluated by cross-comparisons, running the different computer codes for typical test cases. In particular the performances of the models of Mason and Brown (1994), of Andrén and Moeng (1993), of Nieuwstadt and Brost (1986) and of Schmidt and Schumann (1989) have been compared. As regards the case of the convective BL (see Nieuwstadt et al. (1990)) little differences have been found, (within 10%) and it has been attributed to the different values of the Smagorinsky constant used in the codes. Such good agreement could also be due to the fact that the convective boundary layer is very easy to simulate *via* LES, since the turbulence structures are dominated by the large buoyancy-driven scales. Later on, Andrén et al. (1994) have carried out comparative tests among the different models for the case of the neutral stratified planetary Ekman layer. The model of Mason and Thomson (1992) which includes a part that takes into account for local backscatter of

TKE and scalar variance has also been used. All the codes have used similar boundary conditions at the top of the domain and at the bottom boundary. In this case larger differences among the results have been found. In particular, differences have been found both in the mean velocity profile (Fig. 25) and in the second order statistics (Fig. 26). The poor prediction obtained with the Smagorinsky models has not been improved by simply changing the value of the constant and the model with backscatter gave in general better results that that without backscatter. The authors also concluded that the effect of the differences in numerical schemes play a minor role compared with the effect of the different SGS models. The above mentioned tests results from separate researches have clearly shown that the Smagorinsky model gives poor results when complex physics is involved (for example the case of rotation), and that models based on a velocity scale that is determined through the solution of a transport equation for the SGS TKE tend to be more accurate. On the other hand, it is worth noting that the choice of a more complete model, that allows to capture physical aspects related to departure from local isotropy and equilibrium of the small scales, is payed in terms of number of constants to be calibrated, constants needed for the parameterization of the terms of the transport equation of q_{sgs}. As an example, the model used by D.W.Denbo and Skyllingstad (1996) for the study of stratification and rotation effects in the ocean makes use of seven independent constants.

The stably stratified boundary layer is even more difficult to simulate compared with the convective and the neutral counterparts. In stable stratification an additional length scale has to be considered w/N, where w is a characteristic velocity-scale and N is the Brunt-Vaisala frequency. The increase of stable stratification, and, thus of N, reduces the characteristic length-scale of the flow field. Moreover, under strong stratification additional effects may occur, like local relaminarization, low-level jets and internal gravity waves. A successful simulation of the large-scale SBL, under weak stratification has been carried out by Mason and Derbyshire (1990). The authors have solved the Boussinesq form of the governing equations for the case of infinite Reynolds number (meaning that the molecular diffusivity is considered to be negligible when compared to the turbulent one). The authors have used a Smagorinsky-like model, with additional terms that take into account the effects of stratification. The turbulent stresses are expressed as:

$$\tau_{sgs,ij} = 2\nu_{sgs}\overline{S}_{ij}$$

with the SGS eddy viscosity given by:

$$\nu_{sgs} = \frac{\lambda^2|S|}{\phi^2}.$$

The eddy viscosity is thus expressed as the product between a length-scale λ/ϕ and a velocity scale $\lambda|S|/\phi$. The term ϕ takes into account the effects of stratification on the length-scale and subgrid velocity scale:

$$\frac{1}{\phi} = 1 - \beta R_f$$

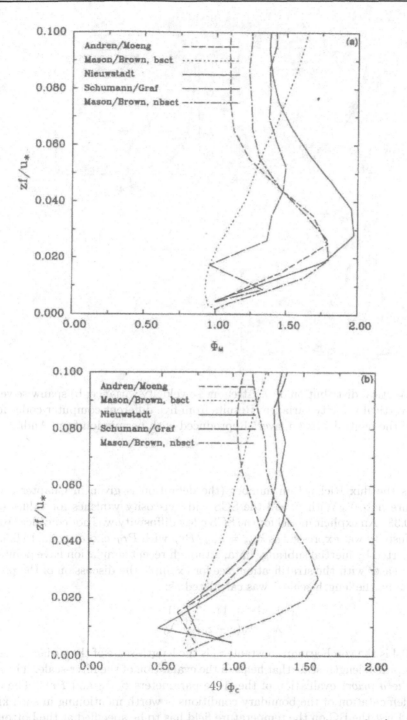

Figure 25. Vertical distribution of the non-dimensional gradient of: a) mean velocity field; b) mean scalar field. Results from five different computer codes for the simulation of the neutral Ekman layer. Reproduced with permission from Andren et al., (1994).

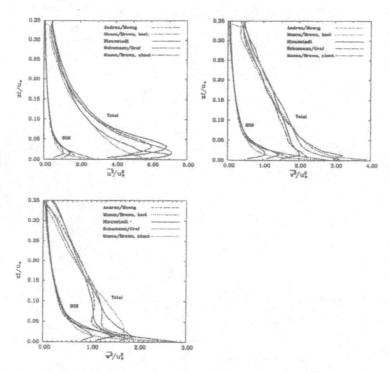

Figure 26. Vertical distribution of: a) streamwise velocity variance; b) spanwise velocity variance; c) vertical velocity variance. Results from five different computer codes for the simulation of the neutral Ekman layer. Reproduced with permission from Andren et al., (1994).

where Ri_f is the flux Richardson number (the definition is given in Chapter 3 of the present lecture notes). With $\beta = 3$ the SGS eddy viscosity vanishes for values of Ri_f larger than 0.33. An explicit model for the SGS eddy diffusivity was not consider. Rather, the eddy diffusivity was expressed as $k_{sgs} = \nu_{sgs}/Pr_T$ with Pr_T chosen equal to 0.5. This value is supported by inertial subrange data, although recent simulation have pointed out that Pr_T increases with the stratification (see for example the discussion of the previous section). Finally, the length-scale λ was calculated as:

$$\frac{1}{\lambda} = \frac{1}{k_v(z + z_0)} + \frac{1}{\lambda_0}$$

where $k_v = 0.4$ is the von Karman constant, z_0 is the length-scale of the surface roughness and λ_0 an empirical length-scale that helps in the evaluation of the filter-scale. The model only needs the *a priori* evaluation of the three parameters β, λ_0 and Pr_T. The choice and the implementation of the boundary conditions is worth mentioning in such kind of simulations. First the BC on the temperature field has to be specified at the bottom and the top boundaries of the domain. Different conditions correspond to different physical situations. It has to be remarked that a simulation of an *actual* atmospheric BL can be hardly performed, due to the variability in time of the heat fluxes. Mason and Derbyshire (1990) have chosen to perform simulations using different BCs for the temperature field. In particular at the top boundary, zero heat flux was considered, whereas at the bottom boundary either constant surface heat flux or a constant surface cooling were considered. As regards the velocity field, zero shear stress (free-slip) was considered at the top boundary. Since the near-wall viscous layer was not resolved, the bottom velocity and temperature were imposed in such a way to simulate the presence of a log-layer in their vertical profiles. The main result of this study was to show the ability of large-scale LES to give qualitative information of the spatial characteristics of the SBL. For example, the analysis of the instantaneous field has shown that the spiral structures, already observed in the neutral case, still survive under stratification (Fig. 27).

The model of Mason and Derbyshire (1990) has proved to work satisfactorily in cases of weak stratification, supporting the local scaling arguments of Nieuwstadt (1986), discussed with details in Chapter 4 of the present lecture notes. For example the simulations have well predicted the decrease of the thickness of the boundary layer and the increase of the angle of the geostrophic wind with stratification. On the other hand, no evidence of internal waves, observed, for example in field and laboratory experiments under strong stratification (see Komori et al. (1983)) has been given by the simulation.

According to Mason and Derbyshire (1990) a drawback of the model has been recognized to be the constant value of the SGS Prandtl number. As already discussed in these notes, this is an acceptable approximation in neutral flow or in subcritical conditions only. In a successive work Brown et al. (1994) have repeated and extended the work of Mason and Derbyshire (1990), using an improved SGS model. The model was based on the original formulation of Mason and Derbyshire (1990) and it included two main modifications: the stability parameter has been modified and the *stochastic backscatter* of Mason and Thomson (1992) has been incorporated. The modification of the stability parameter has allowed the increase of the SGS Prandtl number with the Richardson number, and it has been observed to be beneficial in particular as regards the near-surface turbulent statistics. Moreover, the improved model with backscatter has allowed to obtain better agreement between the computed first-order statistics (mean density gradient and shear) and those measured in field experiments.

The case of moderate to strong stratification in the SBL has been investigated by Saiki et al. (2000). In order to deal with the case of strong stability, they modified the original model of Sullivan et al. (1994), that, in the present case gave unphysical profiles of the turbulent quantities when the BL was subjected to large negative heat flux at the bottom surface (corresponding to rapid cooling of the surface). The drawback was overcome by adding a second part to the model for the SGS heat flux, in a similar fashion as already done for the momentum stresses. The inclusion of the second part in the model for the turbulent density flux prevented from the zero near-surface density flux obtained with the original model under rapid cooling of the bottom surface of the BL. The modified model was able to reproduce most of the physics occurring in strongly stable-stratified flows. In particular, the generation of a low level jet (the mean velocity within the boundary layer is larger than the forcing geostrophic wind) has been observed (Fig. 28), and the presence of large-scale internal waves has been detected.

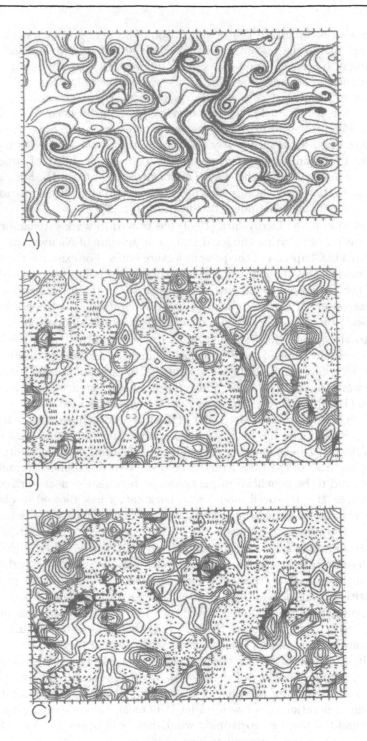

Figure 27. Instantaneous contouring of: a) fluctuating streamwise velocity; b) vertical velocity; temperature; at a level $z = 67.5$ m in case of stably stratified flow. Reproduced with permission from Mason and Derbyshire (1990).

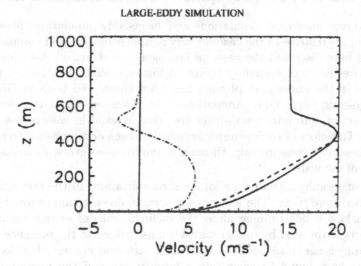

Figure 28. Vertical profiles of velocity. Dashed line, $\langle u \rangle$; dashed-dotted line $\langle v \rangle$; solid line, module of velocity. Reproduced with permission from Saiki et al., (2000).

It is noteworthy that, as already discussed, the dynamic model has been shown to be able to reproduce the salient characteristics of the strongly stratified boundary layer, at least in cases where the wall layer was completely resolved (Armenio and Sarkar (2002). The performance of dynamic eddy-diffusivity models used in conjunction with wall-function or with a separate evaluation of the near-wall layer (for a review see Piomelli and Balaras (2002); Squires (2003)) in order to deal with large Reynolds has been very recently successfully used in engineering applications in cases of neutrally stratified flows. It has to be expected that, in the next future, environmental and geophysical applications, like the study of the atmospheric boundary layer or of the oceanic mixed layer will take advantage of these novel hybrid techniques.

3 numerical studies of topographic effects in stratified flows

The results of direct numerical simulations and large-eddy simulations shown in the previous sections, are relative to the cases of very simple geometry and complex physics. In particular we have discussed the cases of homogeneous stratified flows and those of stratified flows over flat-wall boundary layers. Additional effect of rotation, that leads to complications in the underlying physics, has been considered both in DNS studies and in the large-scale Large eddy simulations. As discussed, the latter were carried out using an *a priori* determined wall-layer law, that is valid in case of not separated boundary layer. The effect of surface imperfections has been accomplished treating them as a wall roughness and, consequently, through a roughness length-scale included in the logarithmic law of the wall.

Large-scale topography introduces additional complications in the flow field, in particular in stratified conditions. The lee-waves generated downstream of an obstacle and the severe downslope winds running along the inclined walls of mountains are typical examples of effects introduced by topography when associated to the presence of stratification. Topography is also responsible of additional turbulent mixing, that, for example, in the ocean has been found to constitute a large fraction of the measurable overall mixing. Analytical theories, laboratory and field experiments in the past were devoted at understanding the role of topography in the evolution of the flow field. A detailed description of these topographic effects are in Baines (1995). Large-scale numerical simulations of the stratified flow over a topography have recently been carried out, among the other, by Legg (2004a,b); Legg and Adcroft (2004), using MITgcm (MIT general circulation model). This code solves the 3D formulation of the Reynolds-averaged Navier Stokes equation with a very crude turbulence model, namely using constant values of the horizontal and vertical eddy viscosities and diffusivity. In such simulations, the full-scale problem was investigated with the cell size of the order of hundred meters in the horizontal direction and tent of meters in the vertical one

A better understanding of small-scale processes can be gained in equivalent laboratory-scale simulations, carried out at small Reynolds number (some order of magnitude smaller than the ocean-scale value) and solving directly all the turbulence scales (via DNS) or the large, energy-carrying one (via resolved LES). In this section we discuss some of these aspects based on very recent results for small-scale simulations.

Armenio et al. (2003) have performed resolved large-eddy simulations of the steady flow over a corrugated wall. The wall was equipped with longitudinal ridges whose height was set equal to $1/8$ the depth of the domain. The friction Reynolds number Re_τ (based on the friction velocity u_τ and fluid depth) was set equal to 400 that, in the neutral case, gave a bulk Reynolds number (based on the mean velocity) roughly equal to 6400. As boundary conditions, the authors have used a no-slip condition over the bottom surface and a free-slip condition at the top-boundary, thus miming the presence of a zero-Froude free surface. Constant perturbation density was considered at the top-bottom boundaries, corresponding to isothermal conditions, and, finally the molecular Prandtl number was set equal to 5, that corresponds to thermally stratified water. The case

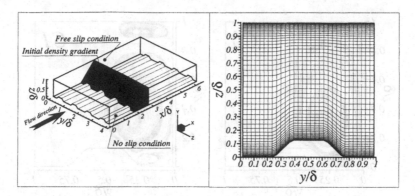

Figure 29. Schematic of the problem studied by Armenio et al. (2003) together with a view of the cell distribution in the cross-stream plane. Reproduced with permission.

of stable stratification was studied. A sketch of the domain configuration investigated is in Fig 29. Several cases of stable stratification were considered, from the neutral flow to moderate stratification. The SGS model of Armenio and Sarkar (2002) was used. The study has evidenced that stable stratification primarily affects the secondary recirculations, that typically arise in along-ridge flow fields, due to the unbalance of the normal Reynolds stresses in the cross-stream planes. In spite of their own weakness (the mean velocity in the secondary flows is of the order of $3 - 5\%$ of the mean streamwise velocity) these recirculations constitute an effective mean for the overall mixing in the fluid column, driving warm (light) fluid from the top to the bottom surface and cold (heavy) fluid from the bottom to the free-surface. Figure 30 shows the response of the secondary recirculations to stable stratification. In case of passive scalar, we can observe the presence of a large scale secondary flow spanning the whole fluid column, and the presence of a small cellular flow at the trough of the ridge. The increase of stratification produces the formation of a strong density gradient in the free surface region that creates a barrier of potential energy and reduces the vertical extension of the main secondary circulation. Conversely, the small cellular flow at the trough of the ridge appears increased with stratification. It has also been observed that the intensity of the recirculations increases with stratification. This was attributed to the fact that stratification increases anisotropy in the flow field, and in particular it reduces the ratio between the vertical turbulence intensity $\langle w'w'\rangle^{1/2}$ and the spanwise one $\langle v'v'\rangle^{1/2}$. Since the production of the mean streamwise vorticity Ω_x is equal to

$$\frac{\partial^2}{\partial y \partial z}(\langle v'v'\rangle - \langle w'w'\rangle),$$

it clearly appears that the increased anisotropy enhances the streamwise mean vorticity and consequently the intensity of the recirculations. The wall stresses also appear to be affected by stratification. Figure 31 shows that the primary wall stress is strongly reduced in the trough of the ridge whereas it appears enhanced over the inclined wall and the top of the ridge. This redistribution of the primary wall stress along the spanwise

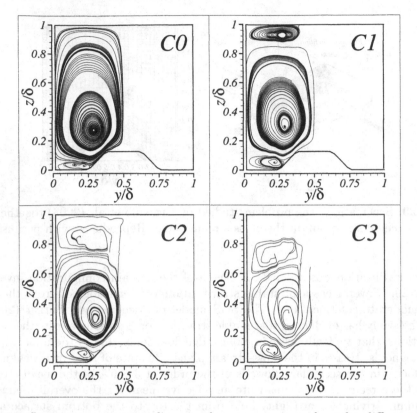

Figure 30. Instantaneous streamlines in the cross stream plane for different levels of stable stratification. Stratification increases from C0 (passive scalar) to C3. Reproduced with permission from Armenio et al. (2003).

direction is due to a corresponding re-distribution of the Reynolds stress $\langle u'w' \rangle$ along y. The secondary wall stress increases with stratification. This effect is related to the increase of the intensity of the small cellular recirculation located at the trough of the ridge.

Summing up, the analysis of the stably stratified along-ridge flow has shown that stratification produces:

- an upper, free-surface region characterized by the presence of a strong density gradient that deepens with increased stratification in the fluid column. In this region vertical mixing of mass and momentum is strongly inhibited and counter-gradient density fluxes are observed.
- an inner region characterized by the presence of secondary recirculations and vertical mixing. The vertical extension of this region decreases with stratification, due to the intrusion of the strong density-gradient in the core of the flow field. However the increased anisotropy of the turbulent field enhances the intensity of the secondary recirculations, and thus the associated large scale mixing.

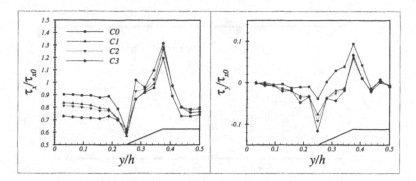

Figure 31. Spanwise distribution of: a) primary wall stress; b) secondary wall stress, for different levels of stable stratification. Stratification increases from C0 (passive scalar) to C3. Reproduced with permission from Armenio et al. (2003).

The effect of stable stratification on the across-ridge flow was studied *via* LES by Calhoun et al. (2001). The authors have considered a large amplitude sinusoidal bottom wall with $2a/L_w = 0.1$ where $2a$ is the wave height and L_w is the wave length. The Reynolds number of the simulation, for the neutral case, was set equal to 7000, based on the mean velocity and the depth of the fluid column. The molecular Prandtl number was chosen equal to one. Two levels of overall stratification were considered, namely $Ri_\tau = 31$ and $Ri_\tau = 62$ with the friction Richardson number defined as

$$Ri_\tau = \frac{gh\Delta\rho}{\rho_0 u_\tau^2},$$

with h the fluid depth and $\Delta\rho$ the density gap between the bottom and the top boundary. Isothermal boundary conditions were considered in this study. Based on the results of Garg et al. (2000) the two levels of stratification analyzed were classified to belong respectively to a *buoyancy affected* regime and to a *buoyancy dominated* regime. The more recent study of Armenio and Sarkar (2002) has shown that, for the Reynolds number used in the simulations of Calhoun et al. (2001) the levels of stratification belong to the case of weakly stratified, subcritical regime.

Calhoun et al. (2001) showed that the mean velocity field appears affected by stratification. In particular, Fig. 32 shows a downstream recirculation region smaller than that observed in the case of neutral flow (not shown here). However, the increase of stratification (compare Fig. 32a with Fig. 32b) has only a little effect on the extent of the recirculation region.

Consistently to other studies of stably stratified turbulent flows, Calhoun et al. (2001) also have found out an increase of turbulent anisotropy in the flow field (Fig. 33). In particular, anisotropy appeared more pronounced over the crest of the topography, whereas it appeared less affected by stratification in the re-recirculation region.

The study of Calhoun et al. (2001) has not evidenced the presence of lee waves downstream the wave crest. This might be attributed to the weakness of overall stratification investigated.

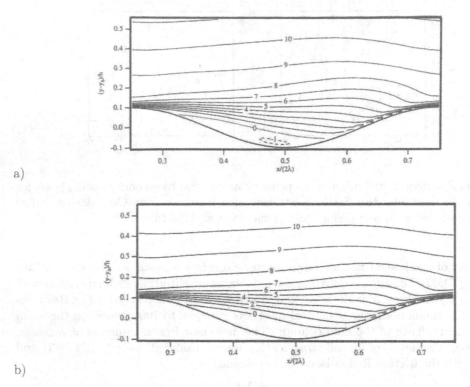

a)

b)

Figure 32. Contours of mean streamwise velocity (made non dimensional with the mean friction velocity): a) Case $Ri_\tau = 31$; b) Case $Ri_\tau = 62$. Reproduced with permission from Calhoun et al. (2001).

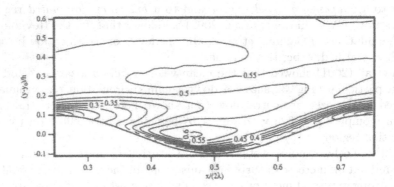

Figure 33. Anisotropy coefficient w_{rms}/u_{rms} in case $Ri_\tau = 31$. Reproduced with permission from Calhoun et al. (2001).

Sullivan and McWilliams (2002) have investigated the interaction between a train of sinusoidal waves and a stratified Couette flow. The study was motivated by the need to gain a better understanding of the interaction between the atmosphere and the ocean, interaction occurring at the air-sea interface. The wave slope was chosen equal to $ak = 0.1$, where $k = 2\pi/L_w$ is the wavenumber. This corresponds to a ratio $a/L_w = 0.015$ for which the streamlines do not separate downstream the crest of the wave. The authors have considered a wide range of levels of stratification, from convective conditions, corresponding to negative values of the Richardson number to stably stratified flow, corresponding to positive values of Ri. The bulk Reynolds number of the simulation was set equal to 8000, that corresponds to values of the friction Reynolds number ranging from 100 to 164 depending on the stratification and the phase speed of the waves. Different values of the phase speed c of the waves were considered, from $c/U_0 = 0$ to $c/U_0 = 0.7$. In the latter case the relative velocity between the air current and the wave crests reaches the minimum value. Isothermal conditions were considered at the top-bottom surfaces of the computational box. An important question, namely the influence of stratification on the pressure field acting over the wave train, was investigated with details. Figure 34 shows how the surface form stress depends on the relative velocity of the waves compared to U_0 and on the stratification. For fast-moving waves, for which the surface form stress is very small or negative, stratification, in term of both sign and overall magnitude, has a negligible effect. On the other hand, for slow-moving waves, $(c/U_0 < 0.5)$ the form stress is significant and it is dramatically affected by stratification. In particular for $c/U_0 = 0.5$ it was observed to decrease by a factor 3 when Ri varied from -0.094 (convective conditions) to 0.031 (stable stratification).

The analysis of a three-dimensional stratified flow field, over an isolated hill was carried out by Ding et al. (2003). As regards the case of the flow over a hill, the flow field is considered to be weakly or strongly stratified depending on the ratio U_0/Nh where h is the height of the hill, U_0 is the far-field velocity and N the Brunt-Vaisala frequency. If this ratio is larger than one, the flow field is considered to be weakly stratified, whereas the opposite is true for values smaller than one. Note that the parameter defined above can be regarded as the ratio between a buoyancy time-scale $1/N$ and an inertial time-scale h/U_0. When the buoyancy time-scale is smaller than the inertial one, the flow appears strongly stratified. In the past, the analysis of the flow field around a hill, under stable stratification has received considerable attention, due to the implications in the dispersion of pollutants. In particular, depending on the location of a source point where a pollutant is released, with respect to the distribution of the mean streamlines, completely different characteristics of dispersion can be observed. Ding et al. (2003) have considered an axis-symmetric hill that has the same shape as that used in a previous experimental study (see Hunt and Snyder (1980)). A laboratory-scale Reynolds number was considered, and Large-eddy simulations were carried out using a dynamic mixed model, and solving directly the wall layer (*resolved* LES). The simulations were carried out for several levels of stratification, for values of U_0/Nh ranging from 1 to 0.2. As regards the response of the flow field to stratification the authors have shown that, consistently with literature results, the increase of stratification suppresses the vertical motion. Figure 35 shows that for $U_0/NH = 0.2$ the streamlines are allowed to flow around the obstacle horizontally,

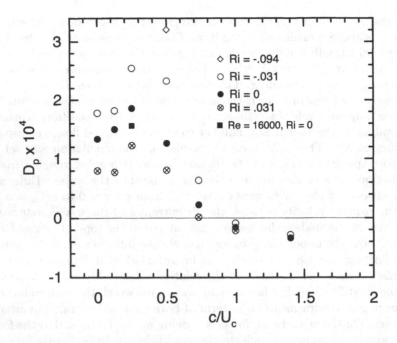

Figure 34. Surface form stress for several values of c/U_0 with varying Ri. Reproduced with permission from Sullivan and McWilliams (2002).

as the vertical rise of the upstream streamline is inhibited by the strong stratification. Since, in these conditions, horizontal motion is predominant, lee-waves are not observed.

As stratification decreases (U_0/Nh increases) vertical motion is allowed for, and the presence of lee-waves is clearly detected (Fig 35). Finally, in case on neutral stratification, a large separation region downstream the hill is observed (Fig. 36). The results of the simulation were in pretty good agreement with those of literature experimental studies (Fig. 36).

The numerical analysis was also used for the verification of analytical theories used for the prediction of the streamline patterns around a hill. In particular the authors have checked the validity of the Sheppard's formula:

$$z_t = h(1 - \frac{U_0}{Nh})$$

that gives the height of the limiting streamwise in the upstream region (z_t) as a function of the stability parameter U_0/Nh. The limiting streamwise is defined as the centerplane streamlines that separates the fluid flowing over the hill from that flowing more horizontally around the hill (Hunt and Snyder (1980)). The decrease of the stability parameter gives $z_t \rightarrow h$ thus indicating horizontal motion. Although the Sheppard's formula was derived for inviscid flow, the numerical results seem to lend justification to its validity, in particular for decreasing values of the stability parameter (Fig 37).

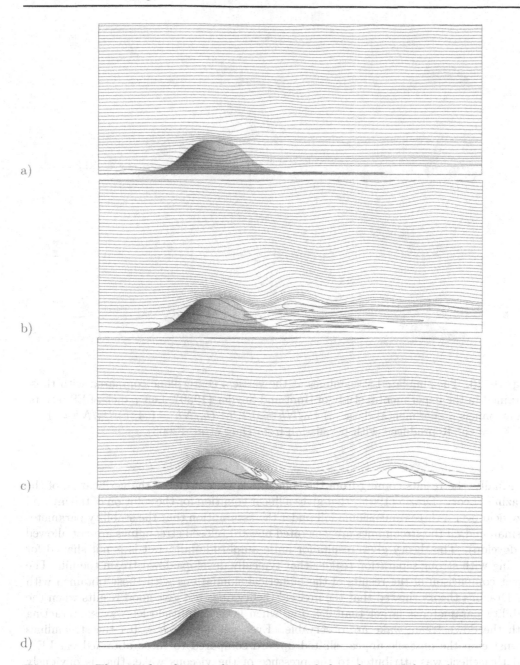

Figure 35. The simulated streamlines in a vertical plane far $0.05h$ from the centerplane: a) $U_0/Nh = 0.2$; b) $U_0/Nh = 0.4$; c) $U_0/Nh = 1$; d) $U_0/Nh = \infty$. Reproduced with permission from Ding et al. (2003).

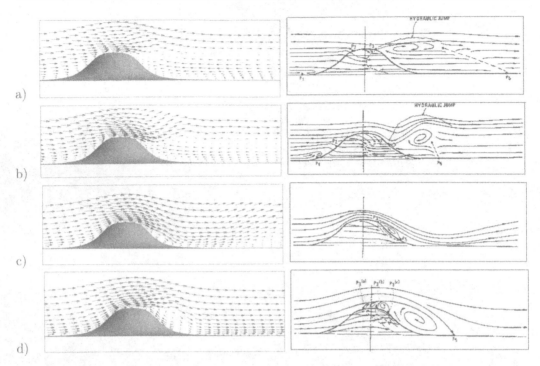

Figure 36. The simulated streamlines at the vertical centerplane compared with those obtained in the experimental study of Hunt and Snyder (1980). Left panels, LES results; Right panels, experimental data. a) $U_0/Nh = 0.2$; b) $U_0/Nh = 0.4$; c) $U_0/Nh = 1$; d) $U_0/Nh = \infty$. Reproduced with permission from Ding et al. (2003).

Finally, the results coming from the simulations were used for the verification of the Drazin's theory (Drazin (1961)) that gives a linear equation for the solution of the inviscid flow field over a three-dimensional hill under the assumption that the stability parameter is small and all the streamlines are generated upward (closed streamlines are not allowed to develop). The theory gives results for finite amplitude hills, but is is not shaped for dealing with strong separation regions that typically develops downstream the hill. The direct comparison of the results of the numerical simulations with those obtained with the Drazin's theory showed that the inviscid theory does not give good results when the stability parameter is larger than 0.4, *i.e.* when large amplitude lee-waves interacting with the downstream wake are observable. Even when $U_0/Nh < 0.4$ the streamlines downstream the obstacle appear shifted upward compared to those obtained *via* LES, and this effect was attributed to the presence of the viscous wake, that is obviously not considered in the inviscid theory. Finally the three-dimensional lee-waves obtained in the numerical simulations were compared with those calculated using the *3D* linear analysis presented in Baines (1995). Although the gross structure of the lee waves can be obtained by the use of the linear theory, Ding et al. (2003) have clearly shown that it underestimated the wave amplitude and overestimate the wave length (Fig. 38a). A

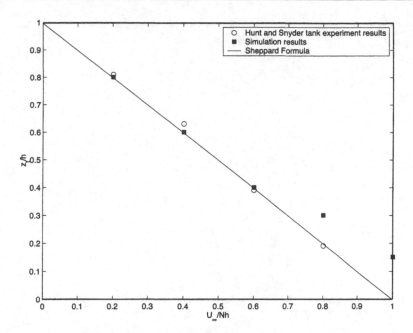

Figure 37. Critical height z_t obtained in the numerical simulations, in the experiments of Hunt and Snyder (1980) and from the analytical Sheppard formula. Reproduced with permission from Ding et al. (2003).

modification of the shape of the obstacle for incorporating the effect of the wake in the linear theory did not improve the results (Fig. 38b).

So far we have discussed investigations of steady currents flowing over a topography, or a wave-train. Moreover, in all cases isothermal conditions (constant perturbation density) were considered. Such conditions, although idealized, better fit the physical characteristics of the atmospheric boundary layer, that is often forced by a mean, steady geostrophic wind and by constant temperature (or constant non-zero heat flux) at the ground surface.

On the other hand, ocean applications are often characterized by two main features:

- the sea bottom is recognized to be adiabatic and thus no-flux conditions are commonly used for the temperature (or, equivalently for the density);

- the driving motion is hardly steady. The most common forcing in the ocean is given by the tides, whose period can be of the order of hours (for example the *M2*-tide has a period slightly larger than 12 hours).

Even when associated to the simplest topography (an inclined wall), the adiabatic condition gives rise to a flow field, due to the inclination of the isopycnals (see Fig. 39). This effect is well known and was first explained by Phillips (1970) through a laboratory experiment.

As regards the effect of different boundary conditions on the characteristics of vertical

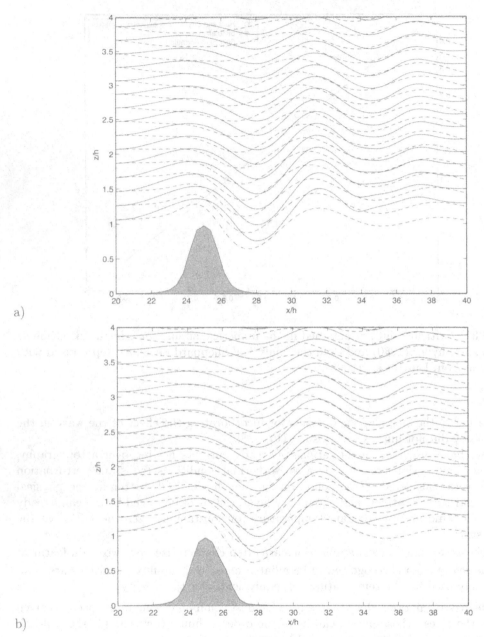

a)

b)

Figure 38. Lee waves obtained by the use of the linear theory (dashed lines) and by large eddy simulation (solid lines) for the stability parameter $U_0/Nh = 1$: a) direct use of the linear theory; b) use of the linear theory over a modified obstacle for taking into account for the effect of the wake. Reproduced with permission from Ding et al. (2003).

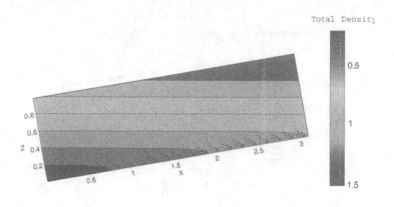

Figure 39. Numerical evidence of the up-slope flow coming from the inclination of the isopycnals over an inclined wall. From Taylor et al. (2004b).

mixing in a stably stratified flow, a very recent paper by Taylor et al. (2004a) shows that a free-surface channel flow, with adiabatic bottom surface and positive heat flux supplied from the top, case prototypical of shallow water sea conditions, behaves differently from the idealized case of a closed channel flow with isothermal conditions at the boundaries already studied by Armenio and Sarkar (2002).

The effect of a tide over a topography, under stable stratification, gives rise to a very interesting effect that was observed to occur in the oceans. In presence of stable stratification, when a barotropic tide flows over a bumpy topography, energy is transferred from the tide to internal gravity waves, that act to damp the tidal energy by radiation. This process is well known (for a review the reader is referred to Wunsch (1975)) and very recently was revisited by Young and co-workers (Llewellyn-Smith and Young (2002, 2003); Balmforth et al. (2002); Petrelis et al. (2004)). A schematic of the physical problem is in Fig. 40.

Bell (1975) has found an analytical estimation of the conversion rate of energy from the tide to the radiated internal waves, under the assumptions that the water column is unbounded aloft, the B-V frequency is uniform along the water depth and the amplitude of topography is very small. Llewellyn-Smith and Young (2002) have extended the results of the Bell's theory removing the underlying assumptions and adding the effect of rotation. The only restriction of their analytical study is on the amplitude of the ridge that has to be considered very small. This assumption derives from the fact that in the

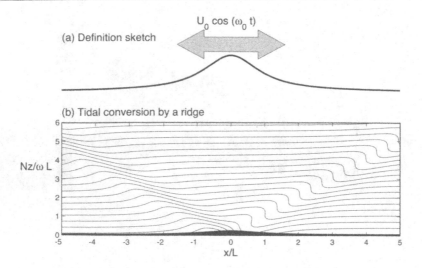

Figure 40. Schematic of the barotropic tide over a bump and representation of the conversion of tidal energy in internal waves. Reproduced with permission from Llewellyn-Smith and Young (2002).

analytical study, the boundary conditions are placed at a flat surface. The authors have found that the conversion rate for a finite-depth ocean can be much smaller than the infinite-depth counterpart, when the length-scale of the topography is comparable with the horizontal wave-length of the internal tide. In a successive work, Balmforth et al. (2002) have estimated the conversion rate of energy from barotropic tide to radiation of internal waves. The fluid column was considered inviscid and unbounded at the top (infinite depth assumption), the effect of rotation was included and different topographies were studied, namely a sinusoidal ripple, a set of Gaussian bumps and a random topography. The main aim of the research was to evaluate the dependence of the conversion rate on the amplitude of the topography, the latter quantified by a non-dimensional parameter ϵ_* that is the ratio between the maximum slope of the topography and the slope of an internal tidal ray. Note that the analytical theory of Bell (1975) holds when $\epsilon_* \ll 1$. In Balmforth et al. (2002) subcritical conditions ($\epsilon_* \leq 1$) were investigated. The characteristics of radiation for two different topographies are illustrated in Fig. 41 and Fig. 42. The most important results of Balmforth et al. (2002) can be summarized as follows:

- for $\epsilon_* \ll 1$ the Bell's theory that gives the conversion rate proportional to h^2 (where h is the height of the topography) is accurate;

- for $0 \leq \epsilon_* < 1$ the conversion rate slightly increases with ϵ_*;

- for $\epsilon_* \to 1$ the increase of the conversion rate is strongly dependent on the shape of topography. For instance, for sinusoidal ripples it amounts to about 56%, for a Gaussian topography the increase is about 14% whereas, for random topography, it increases by only 6% with respect to the Bell's prediction.

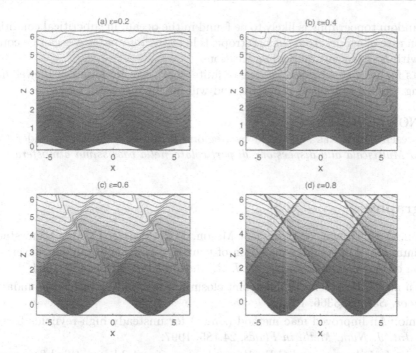

Figure 41. Isopycnals over a sinusoidal ripple for different values of ϵ_*. Reproduced with permission from Balmforth et al. (2002).

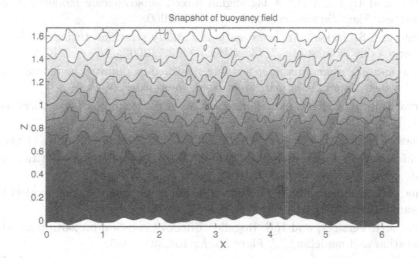

Figure 42. Isopycnal over a random topography. It can be observed that the beams emanate from the steepest slopes of the topography. ($\epsilon_* = 0.6$). Reproduced with permission from Balmforth et al. (2002).

Since a random topography is likely to be found in the ocean, in subcritical conditions the conversion rate of energy from the barotropic tide to the internal waves can be considered to scale with h^2 in most practical applications.

Results for very steep ridges in case of finite depth can be found in Llewellyn-Smith and Young (2003); Petrelis et al. (2004) and will not be discussed here.

ACKNOWLEDGMENTS
The present work was supported by Regione Friuli Venezia Giulia under Contract *Approccio Multiscala alla dispersione di particolato nella bassissima atmosfera*

Bibliography

A. Andrén, A.R. Brown, J. Graf, P.J. Mason, C.H. Moeng, F.T.M. Nieuwstadt, and U. Schumann. Large-eddy simulation of neutrally stratified boundary layer: A comparison of four computer codes. *Q. J. R. Meteorol. Soc.*, 120:1457, 1994.

A. Andrén and C.H. Moeng. Single-point closure in neutrally stratified boundary layer. *J. Atmos. Sci.*, 50:3366, 1993.

V. Armenio. An improved mac method (simac) for unsteady high-reynolds free surface flows. *Int. J. Num. Meth. in Fluids*, 24:185, 1997.

V. Armenio, L. Falcomer, and G.F. Carnevale. Les of a stably stratified flow over longitudinally ridged walls. In *Direct and Large-Eddy simulation V*, page 299. R. Friedrich, B.J. Geurts and O. Metais editors, Kluwer Academic Publishers, 2003.

V. Armenio and U. Piomelli. A lagrangian mixed subgrid-scale model in generalized coordinates. *Flow Turbulence Combust.*, 65:51, 2000.

V. Armenio and S. Sarkar. An investigation of stably stratified turbulent channel flow using large-eddy simulation. *J. Fluid Mech.*, 459:1, 2002.

V. Armenio and S. Sarkar. Mixing in a stably stratified medium by horizontal shear near vertical walls. *Theoret. Comput. Fluid Dyn.*, to appear, 2004.

S.P.S. Arya. Buoyancy effects in an horizontal flat-plane boundary layer. *J. Fluid Mech.*, 68:321, 1975.

P.G. Baines. *Topographic effects in stratified flows*. Cambridge university press, 1995.

N.J. Balmforth, G.R. Ierley, and W.R. Young. Tidal conversion by subcritical topography. *J. Phys. Oceanogr.*, 68:2900, 2002.

J. Bardina, J.H. Ferziger, and W.C. Reynolds. Improved subgrid scale models for large eddy simulation. In *AIAA paper No 80-1357*, 1980.

J. Bardina, J.H. Ferziger, and R.S. Rogallo. Effect of rotation on isotropic turbulence: computation and modelling. *J. Fluid Mech.*, 154:321, 1985.

T.H. Bell. Topographically generated internal waves in the open ocean. *J. Geophys. Res.*, 80:320, 1975.

A.R. Brown, S.H. Derbyshire, and P.J. Mason. Large eddy simulation of stable atmospheric boundary layers with a revised stochastic subgrid model. *Q.J.R. Meteorol.*, 120:1485, 1994.

W. Cabot and P. Moin. Large eddy simulation of scalar transport with the dynamic subgrid-scale model. In *Large Eddy Simulation of Complex Engineering and Geophisical Flows*, page 141. Cambridge University Press, 1993.

R.J. Calhoun and R.L. Street. Turbulent flow over a wavy surface: Neutral case. *J. Geophys. Res-Ocean*, 106:9277, 2001.

R.J. Calhoun, R.L. Street, and J.R. Koseff. Turbulent flow over a wavy surface: Stratified case. *J. Geophys. Res-Ocean*, 106:9295, 2001.

G.N. Coleman, J.H. Ferziger, and P.R. Spalart. A numerical study of the turbulent ekman boundary layer. *J. Fluid Mech.*, 213:313, 1990.

G.N. Coleman, J.H. Ferziger, and P.R. Spalart. A numerical study of the stably stratified turbulent ekman layer. *J. Fluid Mech.*, 244:677, 1992.

S. Corrsin. On the spectrum of isotropic temperature fluctuations in isotropic turbulence. *J. Appl. Phys.*, 22:469–473, 1951.

J.W. Deardoff. The use of subgrid transport equations in a three-dimensional model of atmospheric turbulence. *J. Fluid Engineering, Trans. ASME*, 95:429, 1973.

J.W. Deardoff. Stratocumulus-capped mixed layers derived from a three-dimensional model. *Boundary-Layer Meteorol.*, 18:495, 1980.

L. Ding, R.J. Calhoun, and R.L.V. Street. Numerical simulation of strongly stratified flow over three-dimensional hill. *Bound-Layer Meteor.*, 107:81, 2003.

P.G. Drazin. On the steady flow of a fluid of variable density past an obstacle. *Tellus*, 13:239, 1961.

D.W.Denbo and E.D. Skyllingstad. An ocean large-eddy simulation model with application to deep convection in greenland sea. *J. Geophysical Research*, 101:1095, 1996.

T.M. Eidson. Numerical simulation of turbulent rayleigh-benard convection using subgrid scale modeling. *J. Fluid Mech.*, 158:245, 1985.

E.A. Fadlun, R. Verzicco, P. Orlandi, and J. Mohd-Yusof. Combined immersed-boundary finite-difference methods for three-dimensional complex flow simulations. *J. Comput. Phys.*, 161:35, 2000.

L. Falcomer and V. Armenio. Large-eddy simulation of secondary flow over longitudinally ridged walls. *J. Turbulence*, 3:008, 2002.

R.P. Garg. *Pysics and modeling of stratified turbulent channel flows*. Ph.D thesis, Stanford University, 1996.

R.P. Garg, J.H. Ferziger, and S.G. Monismith. Hybrid spectral finite difference simulations of stratified turbulent flows on distributed memory architectures. *Int. J. Num. Meth. in Fluids*, 24:1129, 1997.

R.P. Garg, J.H. Ferziger, S.G. Monismith, and J.R. Koseff. Stably stratified turbulent channel flows. i. stratification regimes and turbulence suppression mechanism. *Phys. Fluids*, 12:2569, 2000.

M. Germano. The filtering approach. *J. Fluid Mech.*, 238:325, 1992.

M. Germano, U. Piomelli, P. Moin, and W.H. Cabot. A dynamic sub-grid eddy viscosity model. *Phys. Fluids A*, 3:1760, 1991.

T. Gerz, U. Schumann, and S.E. Elgobashi. Direct numerical simulation of stratified homogeneous turbulent shear flows. *J. Fluid Mech.*, 200:563, 1989.

J.R. Herring and O. Metais. Numerical experiments in forced stably stratified turbulence. *J. Fluid Mech.*, 202:97, 1989.

J.O. Hinze. *Turbulence*. McGraw Hill, 1976.

S.E. Holt, J.R. Koseff, and J.H. Ferziger. A numerical study of the evolution and structure of homogeneous stably stratified sheared turbulence. *J. Fluid Mech.*, 237:499, 1992.

J.C.R. Hunt and W.H. Snyder. Experiments on stably and neutrally stratified flow over a model three-dimensional hill. *J. Fluid Mech.*, 96:671, 1980.

F.G. Jacobitz and S. Sarkar. The effect of non-vertical shear on turbulence in a stably stratified medium. *J. Fluid Mech.*, 10:1158, 1998.

F.G. Jacobitz, S. Sarkar, and C.W. Van Atta. Direct numerical simulation of the turbulence evolution in a uniformly sheared and and stably stratified. *J. Fluid Mech.*, 342: 231, 1997.

S.A. Jordan. A large-eddy simulation methodology in generalized coordinates. *J. Comput. Phys.*, 148:322, 1999.

H.-J. Kaltenbach, T. Gerz, and U. Schumann. Large-eddy simulation of homogeneous turbulence and diffusion in stably stratified shear flow. *J. Fluid Mech.*, 280:1, 1994.

L.H. Kantha and C.A. Clayson. *Small scale processes in geophysical fluid flows*. Academic Press, 2000.

J. Kim and P. Moin. Application of a fractional step to incompressible navier-stokes equations. *J. Comp. Phys.*, 59:308, 1985.

J. Kim, P. Moin, and R.D. Moser. Turbulence statistics in fully developed channel flow at low reynolds number. *J. Fluid Mech.*, 177:133, 1987.

Y. Kimura and J.R. Herring. Diffusion in stbly stratified turbulence. *J. Fluid Mech.*, 328:253, 1996.

A.N. Kolmogorov. Energy dissipation in locally isotropic turbulence (in russian). *Dokl. Akad. Nauk. SSSR*, 32:19–21, 1941a.

A.N. Kolmogorov. Local structure of turbulence in an incompressible fluid at very high reynolds numbers (in russian). *Dokl. Akad. Nauk. SSSR*, 30:299–303, 1941b.

S. Komori, H. Ueda, F. Ogino, and T. Mizushina. Turbulent structure in a stably stratified open-channel flow. *J. Fluid Mech.*, 130:13, 1983.

A.G. Kravchenko, P. Moin, and R. Moser. Zonal embedded grids for numerical simulations of wall-bounded turbulent flows. *J. Comput. Phys.*, 127:412, 1996.

B.E. Launder. On the effects of a gravitational field on the turbulence transport of heat and momentum. *J. Fluid Mech.*, 67:569, 1975.

S. Legg. Internal tides generated on a corrugated continental slope. part 1: Cross-slope barotropic forcing. *J. Phys Oceanogr.*, 34:156, 2004a.

S. Legg. Internal tides generated on a corrugated continental slope. part 2: Along-slope barotropic forcing. *J. Phys Oceanogr.*, in press, 2004b.

S. Legg and A.J. Adcroft. Internal wave breaking on concave and convex continental slopes. *J. Phys Oceanogr.*, 33:2224, 2004.

D.K. Lilly. The representation of small scale turbulence in numerical simulation experiments. In *IBM Sci. Comput. Symposium Environ. Sci.*,, page 195. IBM Form 320-195, 1967.

D.K. Lilly. A proposed modification of the germano sub-grid scale closure method. *Phys. Fluids A*, 4:633, 1992.

S.G. Llewellyn-Smith and W.R. Young. Conversion of the barotropic tide. *J. Phys. Oceanogr.*, 32:1554, 2002.

S.G. Llewellyn-Smith and W.R. Young. Tidal conversion at a very steep ridge. *J. Fluid Mech.*, 495:175, 2003.

P.J. Mason and A.R. Brown. The sensitivity of large-eddy simulations of turbulent shear-flow to subgrid models. *Boundary-Layer Meteor.*, 70:133, 1994.

P.J. Mason and S.H. Derbyshire. Large-eddy simulation of the stably-stratified atmospheric boundary layer. *Boundary-Layer Meteor.*, 53:117, 1990.

P.J. Mason and D.J. Thomson. Stochastic backscatter in large-eddy simulations of boundary-layers. *J. Fluid Mech.*, 242:51, 1992.

C. Meneveau, T.S. Lund, and W.H. Cabot. A lagrangian dynamic sub-grid scale model of turbulence. *J. Fluid Mech.*, 319:353, 1996.

F.J. Millero and A. Poisson. International one-athmosphere equation of state of seawater. *Deep Sea Research*, 27A:255–264, 1981.

D.V. Mironov, V.M. Gryanik, C.H. Moeng, D.J. Olbers, and T.H. Warncke. Vertical turbulence structure and second-moment budgets in convection with rotation: A large-eddy simulation study. *Q J Roy. Meteor. Soc.*, 126:477, 2000.

R. Mittal and P. Moin. Suitability of upwind-biased finite difference schemes for large-eddy simulation of turbulent flows. *AIAA J.*, 35:1415, 1997.

C.H. Moeng. A large-eddy-simulation model for the study of planetary boundary-layer turbulence. *J. Atmos. Sci.*, 41:2052, 1984.

C.H. Moeng and P.P. Sullivan. A comparison of shear- and buoyancy-driven planetary boundary layer flows. *J. Atmos. Sci.*, 51:999, 1994.

P. Moin and K. Mahesh. Direct numerical simulation: a tool in turbulence research. *Ann. Rev. Fluid Mech.*, 30:539, 1998.

A.S. Monin and A.M. Yaglom. *Statistical Fluid Mechanics: Mechanics of Turbulence.* MIT Press, Cambridge, MA, 1971.

F.T.M. Nieuwstadt. The turbulent structure of the stable, nocturnl boundary layer. *J. Atmos. Sci.*, 41:2202, 1986.

F.T.M. Nieuwstadt and R.A. Brost. The decay of convective turbulence. *J. Atmos. Sci.*, 43:532, 1986.

F.T.M. Nieuwstadt, P.J. Mason, C.H. Moeng, and U. Schumann. Large-eddy simulation of the convective boundary layer: a comparison of four computer codes. In *Turbulent shear flows 8*, page 313. Springer-Verlag, Berlin, 1990.

A.M. Obukhov. The structure of the temperature field in a turbulent flow. *Izv. Akad. Nauk. SSSR. Ser Geogr. Geophys.*, 13:58, 1949.

I. Orlansky. A simple boundary condition for unbounded hyperbolic flows. *J. Comp. Phys.*, 21:251, 1976.

F. Petrelis, S.G. Llewellyn-Smith, and W.R. Young. Tidal conversion at a submarine ridge. *J. Phys. Oceanogr*, submitted, 2004.

O.M. Phillips. On flows induced by diffusion in a stably stratified fluid. *Deep-Sea Res.*, 17:435, 1970.

U. Piomelli. High reynolds number calculations using the dynamic subgrid-scale stress model. *Phys. Fluids A*, 5:1484, 1993.

U. Piomelli and E. Balaras. Wall-layer models for large-eddy simulations. *Ann. Rev. Fluid Mech.*, 34:349, 2002.

U. Piomelli and J.R. Chasnov. Large-eddy simulations: theory and applications. In *Transition and turbulence modelling*, page 269. edited by D. Henningson, M. Hallbäck, H. Alfreddson and A. Johansson, Kluwer Academic Publishers, Dordrecht, 1996.

M.M. Rai and P. Moin. Direct simulations of turbulent-flow using finite-difference schemes. *J. Comp. Phys.*, 96:15, 1991.

W.C. Reynolds. In *in Whither Turbulence? Turbulence at the Crossroads*, page 313. Springer-Verlag, Heidelberg, 1990.

S.G. Saddoughi and S.V. Veeravalli. Local isotropy in turbulent boundary layers at high reynolds number. *J. Fluid Mech.*, 268:333, 1994.

A.M. Saiki, C.H. Moeng, and P.S. Sullivan. Large-eddy simulation of the stably stratified planetary boundary layer. *Boundary-Layer Meteorol.*, 95:1, 2000.

S. Salon. Turbulent mixing in the gulf of trieste under critical conditions. *Ph.D thesis, Dipartimento di Ingegneria Civile, Universitá di Trieste*, 2004.

H. Schmidt and U. Schumann. Coherent structure of the convective boundary layer derived from large-eddy simulation. *J. Fluid Mech.*, 200:511, 1989.

D.N. Slinn and J.J. Riley. A model for the simulation of turbulent boundary layers in an incompressible stratified flow. *J. Comput. Phys.*, 144:550, 1998.

J.S. Smagorinsky. General circulation experiments with primitive equations.i. the basic experiment. *Mon. Weather Rev.*, 91:99, 1963.

W.D. Smyth and J.N. Moum. Lenght-scales of turbulence in stably stratified mixing layers. *Phys. Fluids*, 12:1327, 2000.

P.R. Spalart and J.H. Wattmuff. Experimental and numerical study of a turbulent boundary-layer with a pressure gradient. *J. Fluid Mech.*, 249:337, 1993.

K.D. Squires. Detached-eddy simulation: current status and perspectives. In *Direct and Large-Eddy simulation V*, page 465. R. Friedrich, B.J. Geurts and O. Metais editors, Kluwer Academic Publishers, 2003.

K.R. Sreenivasan. The passive scalar spectrum and the obukhov-corsin constant. *Phys. Fluids*, 8:189–196, 1996.

C. Staquet. Mixing in a stably stratified shear layer: two- and three-dimensional numerical experiments. *Fluid. Dyn. Res.*, 27:367, 2000.

P.P. Sullivan and J.C. McWilliams. Turbulent flow over water waves in the presence of stratification. *Phys. Fluids*, 14:1182, 2002.

P.P. Sullivan, J.C. McWilliams, and C.H. Moeng. A subgrid-scale model for large-eddy simulation of planetary boundary-layer flows. *Boundary-Layer Meteorol.*, 71: 247, 1994.

P.P. Sullivan, J.C. McWilliams, and C.H. Moeng. A grid-nesting method for large-eddy simulation of planetary boundary-layer flows. *Boundary-Layer Meteorol.*, 80: 167, 1996.

J. Taylor, S. Sarkar, and V. Armenio. Large eddy simulation of stably stratified open channel flow. *submitted*, 2004a.

J. Taylor, S. Sarkar, and V. Armenio. work in progress for les simulation of reflection of internal waves over a topography. 2004b.

H. Tennekes and J.L. Lumley. *A First Course in Turbulence*. The MIT Press, Cambridge, MA, 1972.

J.F. Thompson, Z.U.A. Warsi, and C.W. Mastin. *Numerical grid generation*. North-Holland, 1985.

C.A.G. Webster. An experimental study of turbulence in a density stratified shear flow. *J. Fluid Mech.*, 19:221, 1964.

C. Wunsch. Internal tides in the ocean. *Rev. Geophys.*, 13:167, 1975.

Y. Zang and R.L. Street. Numerical simulation of coastal upwelling and interfacial instability of a rotating and stratified fluid. *J. Fluid Mech.*, 305:47, 1995.

Y. Zang, R.L. Street, and J. Koseff. A non-staggered grid, fractional step method for the time-dependent incompressible navier-stokes equations in curvilinear coordinates. *J. Comp. Phys.*, 114:18, 1994.

Chapter 2:
Internal gravity waves in geophysical fluids

Chantal Staquet

Laboratoire des Ecoulements Géophysiques et Industriels,
Université Joseph Fourier, Grenoble, France
Chantal.Staquet@hmg.inpg.fr

Abstract In this second Chapter, we recall some basic notions on stably-stratified flows before focusing on internal gravity wave dynamics. In Section 1, we illustrate the occurence of stably-stratified flows in nature and, in Section 2, we derive the Boussinesq approximation. Except for a very brief presentation of the Kelvin-Helmholtz instability in Section 3, we start discussing about internal gravity waves from this section on. The linear properties of the wave are discussed in this Section. The mechanisms that can lead the wave field to breaking are addressed in Sections 4 to 6: parametric and buoyancy-induced instabilities (Section 4), interaction with a shear flow (Section 5), interaction with a sloping boundary (Section 6). We briefly discuss about the statistical properties of the breaking wave field in Section 7 and introduce some notions on mixing in Section 8.

1 Stably-stratified fluids: why are they important

1.1 Stably-stratified fluids are ubiquitous

An incompressible fluid is stably-stratified if it displays a vertical density gradient with negative sign (or a vertical temperature gradient with positive sign): the fluid becomes less dense, or warmer, as the altitude increases. Such fluids are ubiquitous in nature: the oceanic water masses, the stratosphere (which is the part of the atmosphere comprised between $\simeq$ 10 and 50 kms) are stably-stratified. This is exemplified in the oceanic case in Figure 1: the temperature profile measured in the Mindanao trench during the 1929-1930 Snellius expedition in eastern Indonesia is displayed as a function of depth. The gradient is positive down to 3500m but compressibility effects become important below this depth leading to a slight warming of the water masses. Such a change of sign in the temperature gradient therefore does not imply that the fluid is unstably stratified but simply results from compression of the water columns. The concept of potential temperature, which is discussed in detail in Chapter 4 of the lecture notes, is introduced to account for compressibility effects. Indeed, the vertical gradient of the potential temperature field is now everywhere positive.

In this example, the huge depth of the temperature record, close to 10 km, along with the sparsity of the sampling, implies that the overall (large scale) behaviour of the temperature profile is captured. If one considers much smaller scales however, the vertical potential temperature gradient is not always positive. An example is provided in the stratosphere in Figure 2, which displays the potential temperature as a function of altitude. While the gradient is positive in overall, negative gradients exist locally, which are due to turbulent motions. In the stratosphere, the temperature and velocity fluctuations are mostly due to internal gravity waves so that this turbulence results from the breaking of the waves, as we shall further discuss it.

Stably-stratified media are also found in the interior of stars. In the Sun for instance, a stably stratified region exists below the convective zone, in the so-called radiative zone. As we shall show it, internal gravity waves can propagate in a stably-stratified medium. It has been proposed by Schatzman (1996) that the superposition of these waves results in a net transport of mass toward the interior of the Sun, which would account for the under-abundance of Lithium observed in this star: the Lithium would be transported

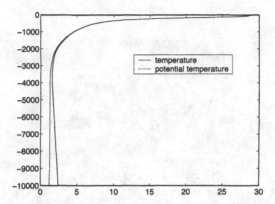

Figure 1. Temperature and potential temperature profiles from data in the Mindanao trench collected during the 1929-1930 Snellius expedition around Indonesia.

toward the core where it would be burned.

Stably-stratified media may also be used within an industrial, or environmental, con-

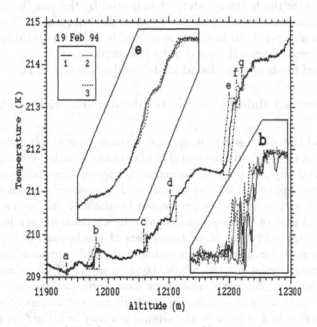

Figure 2. High resolution (20cm) temperature profile, just above the tropopause. Boxes marked with letters (a ...g) contain examples of sheets. Two close-ups of sheets "b" and "e" are displayed with the three profiles measured on the main gondola. Ongoing mixing is clearly present in the sheet "b" (from Dalaudier et al., 1994).

Figure 3. A solar pond (from www.solarpond.utep.edu).

text. The solar pond (Figure 3) is a nice illustration. The idea is to capture heat by a stable stratification. The principle is very simple: the bottom of a pond is made dark and very salty, in order for the bottom water, when heated by the sun (more precisely, when heated by the dark bottom) to remain heavier than the layer above. The temperature of the bottom water may reach 70 to 100 degrees and is used, for instance, to warm water inside a pipe that would cross the pond in its lowest part.

Stably-stratified fluids are also found in the cooling circuits of nuclear reactors.

1.2 Stably-stratified fluids give rise to phenomena with environmental implications

The solar pond just discussed provides one example that stably-stratified fluids may give rise to phenomena with environmental implications. Another common phenomenon, which has a negative impact on the environment, is the well-known situation of a thermal inversion in the atmosphere of a valley in winter. The process is again simple (f.i. Stull, 1988): the ground cools when no longer heated by the sun and, when its temperature becomes lower than that of the neighbouring air, the air cools in turn in transfering heat to the ground. A very cold layer of air thus exists at the bottom of the valley, which is surmounted by warmer air, leading to a stable (potential) temperature gradient. This stable situation may persist for the whole day in winter. If cities are located at the bottom of the valley, the pollutants emitted by road traffic and factories remain trapped in the stably-stratified layer of air, leading to high concentrations of pollutants. Two examples are provided in Figure 4, in the Grenoble valley and the Chamonix valley.

1.3 Quantification of the stability: the Brunt-Väisälä frequency

The Brunt-Väisälä frequency is the frequency of a fluid particle displaced adiabatically from its equilibrium position along the vertical. The demonstration is very standard. Let

Figure 4. (a) The valley of Grenoble in winter, when a thermal inversion exists all day long. (b) Photograph of the Chamonix valley, with the Green Needle (Aiguille Verte) in background. Water and pollutant particles are again trapped in a stable thermal layer at the bottom of the valley.

us do a thought experiment and consider one fluid particle of unit volume, at equilibrium position z in a stably-stratified fluid with density profile $\bar{\rho}(z)$. When the particle is displaced adiabatically by a vertical distance ζ, the local density at the new position is $\bar{\rho}(z + \zeta)$. The fluid particle is thus subjected to two forces, its weight $\bar{g}\bar{\rho}(z)$ and the Archimedean force $-\bar{g}\bar{\rho}(z + \zeta)$. Applying Newton's law to the fluid particle results in the equation $\ddot{\zeta} + N^2\zeta = 0$, where $N^2 = -\dfrac{g}{\bar{\rho}(z)}\dfrac{d\bar{\rho}}{dz}$ is the Brunt-Väisälä, or buoyancy, frequency. Stability implies that N^2 is positive.

A standard approximation to describe stably-stratified incompressible fluid motions is the Boussinesq approximation, described in Section 2.2; in this case, the density and the temperature profiles are linearly related. Therefore, the Brunt-Väisälä frequency may also be defined as $N^2 = \dfrac{g}{\bar{T}(z)}\dfrac{d\bar{T}}{dz}$. When the fluid is compressible, the potential temperature is used instead of the temperature and the expression of the Brunt-Väisälä frequency is the same as the previous one with the temperature profile being simply replaced by the potential temperature profile.

Figure 5. Atmospheric lee waves generated by a wind blowing over two islands.

1.4   What are the generic motions?

As announced above, the generic fluid motions in a stably-stratified fluid are internal gravity waves. Indeed, the sum of the weight and of the Archimedean force acting on a fluid particle displaced from equilibrium position is (at first order in ζ) $-\vec{g}\zeta\frac{d\overline{\rho}}{dz}$, which is a restoring force. This is the restoring buoyancy force. A restoring force in a continuous medium generates wave motions. We shall show in Section 3 that linear waves are indeed a solution of the equations of motions. Before doing so, we provide two illustrations of internal gravity waves in the atmosphere (Figure 5) and in the ocean (Figure 6).

In the atmosphere, the main mechanims for gravity wave generation is the wind blowing over topography. The generated waves are called lee waves and are made visible by clouds. Figure 5 thus displays the cloud cover over the sea when the blowing wind interacts with two islands. The island located on the left of the picture is too high for the air advected by the wind to go over the island and the advected air simply flows around the island. In this case, the fluid particles are not lifted and no wave is generated. Instead, a quasi-two-dimensional pattern (in a horizontal plane) of vortices are produced, of the Karmann street type. By contrast, the island located on the right of the picture is low enough for the advected air to go over the island and a quite different pattern is observed: in this case, waves are generated, with horizontal wave length of the order of the island size.

In the ocean, the two main mechanisms for the generation of internal waves are the wind and the tide. The wind forcing is indirect: the wind mixes the upper layer of the ocean (over a depth of order 100m) and motions at the bottom of this layer perturbs the stably-stratified fluid below which generates internal gravity waves. By contrast, the tide generates internal gravity waves directly, through its interaction with topography (seamounts, continental slopes, etc.) Internal gravity waves generated in the neighbourhood of a continental slope by the tide are displayed in Figure 6. The waves are in the interior of the water masses but, because their amplitude is large and they propagate not far from the surface, the velocity field they induce perturbs the free surface which makes them visible using a radar on board of a satellite.

Figure 6. Internal waves (propagating *below* the free surface) generated by the interaction of the tide with a continental slope. This is a radar image acquired by the Synthetic Radar Aperture aboard of a satellite.

1.5   What about unstably stratified flows?

Unstably stratified flows are also very common in natural media. In this situation, no waves are generated since the buoyancy force is not a restoring one. On the opposite, this force lifts the fluid particle further away from the position it has been brought. The commonest example occurs in the atmospheric boundary layer : the ground is heated by the Sun during the day, which heats the air (the direct heating of air by the Sun is much less important, especially if the air is dry). The vertical gradient of the potential temperature profile is therefore negative in this case. Such a situation is highly unstable and gives rise to large scale turbulent motions, which mix the fluid. Therefore, if pollutants are emitted at the ground level, the local concentrations are generally smaller than in a stable situation, but may be found very far from their emission region. For instance, in summer in the Chamonix valley, high concentrations of heavy metals (such as Mercury) have been found at 4300m altitude, at Dôme du Goûter (Veysseyre et al., 2000).

2 Basic equations

2.1   Governing equations

Let (x, y, z) be a cartesian coordinate system in an absolute reference frame where the fluid velocity is $\vec{u} = (u, v, w)$. The fluid motions in this reference frame are governed by the compressible Navier-Stokes equations. These equations are composed of

• the momentum equation, which is, when expressed per unit mass:

$$\frac{D\vec{u}}{Dt} = -\nabla p + \vec{g} - 2\vec{\Omega} \wedge \vec{u} + \frac{\mu}{\rho}\nabla^2 \vec{u}, \tag{2.1}$$

where $\dfrac{D}{Dt} = \dfrac{\partial}{\partial t} + \vec{u}.\nabla$ refers to the material derivative, p is the pressure field, $-2\vec{\Omega} \wedge \vec{u}$ is the Coriolis acceleration due to earth rotation and μ is the dynamical viscosity;

- the mass conservation equation:

$$\frac{\partial \rho}{\partial t} + \nabla.(\rho \vec{u}) = 0, \tag{2.2}$$

where ρ is the density field;

- and a dynamical equation for the internal energy e, stemming from the first law of thermodynamics applied to a system made of a unit mass of volume $v = 1/\rho$:

$$\frac{De}{Dt} = Q - p\frac{Dv}{Dt}. \tag{2.3}$$

Q is the amount of heat added to the system v and $-p\frac{Dv}{Dt}$ is the work of pressure forces on the system.

2.2   The Boussinesq approximation

Water is a very weakly compressible fluid so that oceanic fluid motions may be assumed to be incompressible over the typical vertical scale of the motions. The fact that density, that is the mass of the fluid system, still varies in space and time may be incompatible with the assumption that the fluid velocity is incompressible. A consistent, and very well-known, approximation that satisfies the former requirement while assuming that the fluid is close to incompressiblity is the Boussinesq approximation. We now describe in detail this approximation (see also Cushman-Roisin, 1994, p. 37).

In the following, we set $\rho = \rho_0 + \tilde{\rho}$ where ρ_0 is some constant reference density and $\tilde{\rho}$ refers to the deviation field about ρ_0. In the ocean for instance, ρ_0 is the mean value of the density, equal to $1028 kg/m^3$ (at atmospheric pressure) and $\tilde{\rho}/\rho_0 \simeq 10^{-3}$. We also introduce reference values for the temperature and the salinity: $T_0 = 10^o C$ and $S_0 = 34 g/kg$.

How does the mass conservation equation become $\nabla.\vec{u} = 0$?   Let us find in which sense the mass conservation equation can be approximated by $\nabla.\vec{u} = 0$.

We introduce the small parameter ϵ such that $\tilde{\rho}/\rho_0 = \epsilon \, \rho^{(1)}/\rho_0$, with $\rho^{(1)}/\rho_0 = O(1)$. We also decompose $\vec{u}$ into $\vec{u}^{(0)} + \epsilon \vec{u}^{(1)}$; hence, $\vec{u}^{(0)}$ is the velocity field in the limit $\epsilon \to 0$, that is, in the limit of infinitely small density fluctuations. Introducing this decomposition of the fields into the mass conservation equation, one gets

$$\rho_0 \nabla.\vec{u}^{(0)} + \epsilon \left(\frac{D}{Dt}\rho^{(1)} + \rho_0 \nabla.\vec{u}^{(1)} + \rho^{(1)}\nabla.\vec{u} \right) = 0. \tag{2.4}$$

At order zero in ϵ, the mass conservation reduces to $\nabla.\vec{u}^{(0)}$. This implies that the momentum equations for an incompressible fluid are also valid in this limit (that is, with $\vec{u} \simeq \vec{u}^{(0)}$). This also implies that the gravity acceleration g should be large enough for the product $g\tilde{\rho}/\rho_0$ to stay finite.

How does the energy equation become the usual heat equation? For any constant volume transformation, the internal energy e is equal to $C_v T$ where C_v is the heat capacity per unit mass and T is the temperature. Equation (2.3) can thus be rewritten as

$$C_v \frac{DT}{Dt} = Q + \frac{p}{\rho^2} \frac{D\rho}{Dt}. \tag{2.5}$$

Since the state variable T has been introduced, we need an equation of state. In the limit of very small density deviation about a constant reference density, the fluid may be assumed to be incompressible, as just seen. Hence, ρ only weakly depends on pressure. Density fluctuations arise from temperature and salinity changes about the background state so that, at first approximation, a linear equation of state may be assumed

$$\rho = \rho_0[1 - \alpha(T - T_0) + \beta(S - S_0)], \tag{2.6}$$

where α is the thermal expansion coefficient and β is the haline contraction coefficient.

Assuming that the heat added to the fluid volume v results from diffusion process, Q can be related to T using Fourier law: $\rho Q = \kappa \nabla^2 T$. Replacing Q by this expression in equation (2.5) and using $\frac{D\rho}{Dt} = -\rho \nabla.\vec{u}$, one gets ($k$ is the thermal conductivity of the fluid)

$$\rho C_v \frac{DT}{Dt} + p\nabla.\vec{u} = k\nabla^2 T. \tag{2.7}$$

In the limit $\tilde{\rho} \ll \rho_0$, that is, $\vec{u} \simeq \vec{u}^{(0)}$ one eventually recover the usual heat equation

$$\frac{DT}{Dt} = \frac{k}{\rho_0 C_v} \nabla^2 T. \tag{2.8}$$

The coefficient $\frac{k}{\rho_0 C_v}$ is the thermal diffusity in the fluid, usually denoted κ. If the density fluctuation $\tilde{\rho}$ mainly arises from temperature fluctuations, the above equations become, using (2.6):

$$\frac{D\tilde{\rho}}{Dt} = \kappa \nabla^2 \tilde{\rho}. \tag{2.9}$$

And what about the momentum equation? We use the assumptions $\tilde{\rho} \ll \rho_0$ and $\rho \simeq \rho_0$. In the momentum equations, ρ comes into play in two terms of quite different physical meaning: (i) in the acceleration and Coriolis terms, $\rho D\vec{u}/Dt$ and $-2\rho \vec{\Omega} \wedge \vec{u}$, and (ii) in the gravity term $g\rho$. While ρ can be replaced by ρ_0 in the former two terms, it certainly cannot be replaced also in the latter term because the fluid would be homogeneous otherwise, with constant density ρ_0. Therefore

- $\rho D\vec{u}/Dt \rightarrow \rho_0 D\vec{u}/Dt$; $-2\rho \vec{\Omega} \wedge u \rightarrow -2\rho_0 \vec{\Omega} \wedge u$.
- $-\nabla p + \rho \vec{g} \rightarrow -\nabla \tilde{p} + \tilde{\rho} \vec{g}$, decomposing the dynamical pressure p into P_0 (hydrostatic pressure) plus $\tilde{p}$ and using the hydrostatic balance of the rest state $dP_0/dz = -\rho_0 g$.

The equations of motions become

$$\rho_0 \frac{D\vec{u}}{dt} = -\nabla \tilde{p} + \tilde{\rho}\, \vec{g} - 2\rho_0 \vec{\Omega} \wedge u + \mu\nabla^2 \vec{u}. \tag{2.10}$$

The Boussinesq equations are equations (2.9) and (2.10) above along with the incompressibility condition $\nabla.\vec{u} = 0$. We recall that this approximation is strictly valid in the limit $\vec{u} = \vec{u}^{(0)}$ and $\tilde{\rho} \ll \rho_0$.

2.3   Non-dimensional parameters

Models of geophysical (or astrophysical) flows are often designed in the laboratory. The idea is to model not the whole system but only a part of it, whereof dynamics are to be studied. For instance, the formation and dynamics of the Great Red Spot of Jupiter has been modelled in a laboratory experiment using two counter-rotating disks, in order to reproduce the zonal shear that prevails on this planet (as in any strongly rotating planet) and controls the dynamics. The scales are very different in the natural and laboratory systems but comparison can be performed by considering the non-dimensional equations of motions. These non-dimensional equations involve a few non-dimensional parameters and the application of the laboratory results to the natural system can be done if the non-dimensional parameters are the same. The other interest of working with non-dimensional equations, especially when natural systems are to be studied, is that one deals with values of order 1. Such a non-dimensionalization is actually necessary when the equations of motions are solved on a computer.

We now list the main non-dimensional parameters that govern the dynamics of a Boussinesq flow (that is, a flow whose dynamics can be described by the Navier-Stokes equations in the Boussinesq approximation) subjected to the Coriolis force.

Let U be the scale of the velocity field, L be a typical length scale of the fluid motions and τ the time scale. Three non-dimensional parameters compare the nonlinear time scale L/U to the time scale τ, depending upon the physical process whereof τ is the time scale: (i) if τ is the time scale of viscous effects, then $\tau = L^2/\nu$ and $Re = \dfrac{\tau}{L/U} = \dfrac{UL}{\nu}$ is the Reynolds number; (ii) if τ is the time scale of stratification effects, then $\tau = N^{-1}$ and $Fr = \dfrac{\tau}{L/U} = \dfrac{U}{NL}$ is the Froude number; (iii) if τ is the time scale of rotation (or Coriolis) effects, then $\tau = (2\Omega)^{-1}$ and $Ro = \dfrac{\tau}{L/U} = \dfrac{U}{2\Omega L}$ is the Rossby number. Finally, the Prandtl number compares the viscosity to the thermal diffusivity: $Pr = \dfrac{\nu}{\kappa}$.

A few other non-dimensional parameters can be inferred from these four ones. These are for instance the Ekman number, which compares viscous and Coriolis effects: $Ek = \dfrac{\nu}{2\Omega L^2} = ReRo$; and the ratio of rotation to stratification effects (also referred to as the Prandtl ratio): $\dfrac{N}{2\Omega} = \dfrac{Ro}{Fr}$.

2.4   Linearization about a basic state

Since we are interested in motions of small amplitude relative to a basic state, it is useful to decompose the fields into a part associated with the basic state and a part associated with the small amplitude motions. Thus we write: $\vec{u} = \vec{U} + \vec{u}'$, $\tilde{\rho} = R + \rho'$, $\tilde{p} = P + p'$ where $\vec{U}$ and R may depend upon the spatial coordinates but (this is our assumption) do not depend upon time, R and P are in hydrostatic balance, and the amplitude of $\vec{u}'$, ρ' and p' is much smaller than that of U, R and P respectively. Introducing this decomposition into the Boussinesq equations and linearizing about the basic state yield the linearized Boussinesq equations. We shall consider two specific cases.

• **The basic state is a fluid at rest in hydrostatic balance.**

In this case, $U = 0$ and, therefore, the amplitude of $\vec{u}'$ is much smaller than 1. The linearized Boussinesq equations become in this case (ν is the kinematic viscosity):

$$\frac{\partial \vec{u}'}{\partial t} = -\frac{1}{\rho_0}\nabla p' - g\frac{\rho'}{\rho_0}\vec{i}_z - 2\vec{\Omega}\wedge u' + \nu\nabla^2\vec{u}' \tag{2.11}$$

$$\frac{\partial \rho'}{\partial t} = -w'\frac{dR}{dz} + \kappa\nabla^2\rho' \tag{2.12}$$

$$\nabla.\vec{u}' = 0. \tag{2.13}$$

• **The basic state is a parallel shear flow $U(z)$ with density field $R(z)$ in hydrostatic balance.**

The linearized Boussinesq equations of motions become in this case

$$\frac{\partial u'}{\partial t} + U\frac{\partial u'}{\partial x} = -w'\frac{dU}{dz} - \frac{1}{\rho_0}\frac{\partial p'}{\partial x} + 2\Omega v' + \nu\nabla^2 u' \tag{2.14}$$

$$\frac{\partial v'}{\partial t} + U\frac{\partial v'}{\partial x} = -\frac{1}{\rho_0}\frac{\partial p'}{\partial y} - 2\Omega u' + \nu\nabla^2 v' \tag{2.15}$$

$$\frac{\partial w'}{\partial t} + U\frac{\partial w'}{\partial x} = -\frac{1}{\rho_0}\frac{\partial p'}{\partial z} - g\frac{\rho'}{\rho_0} + \nu\nabla^2 w' \tag{2.16}$$

$$\frac{\partial \rho'}{\partial t} + U\frac{\partial \rho'}{\partial x} = -w'\frac{dR}{dz} + \kappa\nabla^2\rho' \tag{2.17}$$

$$\nabla.\vec{u}' = 0. \tag{2.18}$$

3 Solutions of the linearized Boussinesq equations: internal gravity waves, Kelvin-Helmholtz instability

3.1 Generalities on linear waves

A linear wave in a homogeneous and steady medium[1] is characterized by a frequency Ω and a wave vector $\vec{k}$. For a plane wave in an infinite medium, the motion is of the form

$$A\, e^{I(\vec{k}.\vec{x}-\Omega t)}, \tag{3.1}$$

---

[1]The properties of the medium remain unchanged by translation of the reference frame and do not depend upon time.

where A is the wave amplitude. The properties of the wave are completely set by the dispersion relation, namely a relation between Ω and $\vec{k}$:

$$\Omega = F(\vec{k}), \tag{3.2}$$

where the function F also involves the fluid parameters upon which the wave propagation depends (N, g, etc.) The dispersion relation does not depend upon space and time because of the homogeneity and steadiness assumptions.

Since the dispersion relation characterizes the wave field, its derivation is an essential step in the wave study. One can distinguish two methods for this purpose. In the classical method, one first needs to derive the linearized equations of motion. Since the medium is homogeneous, the coefficients of these equations are constant. The general form of the wave solution is then substituted in these equations, from where the dispersion relation is obtained.

A general method that gives access directly to the dispersion relation, without using the equations of motion, can be used instead (Lighthill, 1978). Because it is simpler than the classical method and very powerful, we shall rely on this second method to compute the dispersion relation.

3.2   A general method to compute the dispersion relation

The frequency Ω of an oscillating system, whether consisting of a finite number of degrees of freedom (like the pendulum and or chain of springs) or evolving in a continuous medium, can be computed directly in three different ways (Lighthill, 1978):

1. From the restoring force. The square of Ω is then obtained as follows:

$$\Omega^2 = \frac{\text{restoring force}}{\text{displacement x mass}}; \tag{3.3}$$

2. From energetic considerations. Ω^2 is then defined as:

$$\Omega^2 = \frac{\text{generalized stiffness}}{\text{generalized inertia}}, \tag{3.4}$$

where the *generalized stiffness* is the coefficient of the displacement squared in the expression of the potential energy and the *generalized inertia* is the coefficient of the temporal derivative of the displacement squared in the expression of the kinetic energy;

3. From dimensional analysis.

Before applying this general method to internal gravity waves, let us show how it works for a simple discrete system. For this purpose, we consider the harmonic oscillator, whose frequency is $\Omega = \sqrt{\dfrac{K}{m}}$, where K is the stiffness of the spring and m the mass of the object attached to the spring.

In case 1., the restoring force is $-K.x$ where x is the displacement of the mass from equilibrium position. Hence, Ω^2 should be equal to $|K.x|$ divided by $|x|$ times m, which yields the actual expression for Ω. In case 2., the potential energy is $0.5Kx^2$ so that the "generalized" stiffness is K and the kinetic energy is $0.5m\dot{x}^2$ so that the generalized inertia is m. Finally, in case 3., let us assume that Ω is a function of K and m, of the form $K^\alpha m^\beta$. By writing that the unit of Ω, m and K are rad.s^{-1}, kg (also Newton.m^{-1}.s^2) and Newton.m^{-1} respectively, one gets $\alpha = -\beta = 1/2$.

3.3 Internal gravity waves in an infinite medium

The dispersion relation from general principles. In the following, we assume that the stably-stratified medium is homogeneous, which implies that N is constant. We shall assume that the medium is steady throughout the paper.

To compute Ω, we shall use the second expression provided above (case 2.). We first need to compute the potential energy to infer the generalized stiffness. The potential energy is the work done by the restoring force from the current position to the equilibrium position. We showed in section 1.4 that the restoring force is $g\dfrac{d\overline{\rho}}{dz}\zeta$ so that

$$E_p = \int_\zeta^0 g\frac{d\overline{\rho}}{dz}z\,dz = -N^2\rho_0 \int_\zeta^0 z\,dz = 0.5\rho_0 N^2\zeta^2.$$ The generalized stiffness is therefore

$\rho_0 N^2$. Let us now compute the kinetic energy to get the generalized inertia. As we show it in the next section, the motion induced by a monochromatic internal gravity wave occurs in a two-dimensional vertical plane, made by $\vec{k}$ and $\vec{g}$ (this is the propagation plane of the wave). We assume that this plane coincides with the (x, z) plane. Hence, the kinetic energy is $E_k = 0.5\rho_0(u^2 + w^2)$. From continuity equation: $k_x u + k_z w = 0$, implying that $u = -\tan\theta\,w$, where θ is the angle of the wave vector with the horizontal. Since, $w = \dot{\zeta}$, one gets $E_k = 0.5\rho_0(1 + \tan^2\theta)\dot{\zeta}^2 = 0.5\rho_0\dfrac{\dot{\zeta}^2}{\cos^2\theta}$. The generalized inertia is therefore $\dfrac{\rho_0}{\cos^2\theta}$.

Hence, the frequency (and dispersion relation) of a monochromatic internal gravity waves in a homogeneous medium is

$$\Omega^2 = N^2\cos^2\theta. \tag{3.5}$$

It follows that internal gravity waves are *dispersive* waves, because the phase velocity $c = \Omega/|\vec{k}|$ depends upon $\vec{k}$, with an *anisotropic* dispersion relation, because Ω depends upon the angle that $\vec{k}$ makes with the horizontal.

Main properties of internal gravity waves. The properties of internal gravity waves are described in several textbooks (Lighthill, 1978; Leblond and Mysak, 1978; Cushman-Roisin, 1994) and we refer to these books for a detailed account. We make a brief summary of the important properties which we illustrate with laboratory experiments.

We cannot avoid discussing the beautiful experiment of Mowbray and Rarity (1967), which was designed to check the dispersion relation (3.5). A horizontal cylinder oscillates vertically at a constant frequency Ω in a constant N fluid. The vertical motion of the cylinder perturbs the flow and relation (3.5) show that, if $\Omega^2 \leq N^2$, waves will be produced. Linear waves in a non dissipative medium carry the momentum and the

Figure 7. Laboratory experiments performed by Mowbray and Rarity (1967). Visualization is made in a vertical plane and the dark vertical bar is the cylinder support.

energy of their source without material transport (f.i. Andrews et al., 1987). The energy is transported at the group velocity $\partial\Omega/\partial\vec{k}$. For internal gravity waves, the group velocity is perpendicular to $\vec{k}$. Hence the energy should be carried away by the waves along directions making an angle θ with the vertical. This peculiar pattern is displayed in Figure 7 and is referred to as the St-Andrews cross.

Perhaps the most important property of internal gravity waves in a homogeneous stratified medium is that only a time scale is imposed, namely N^{-1}. There is no scale selection by the stratified medium[2]. It follows that in laboratory experiments of internal gravity waves forced by a paddle, the size of the paddle (or, more generally, of the forcing mechanism) selects the scale (f.i. McEwan, 1971). In the experiment by Mowbray and Rarity discussed above, the diameter of the cylinder selects the width of the energetic region. It may thus be expected that in experiments where the forcing mechanism at frequency Ω has no length scale, all wave vectors which make an angle $\cos^{-1}(\Omega/N)$ with the horizontal will be excited, acording to the dispersion relation (3.5). In the laboratory experiments of McEwan and Robinson (1975) and Benielli and Sommeria (1998), internal gravity waves were forced by parametric instability: a tank stably-stratified with salted water oscillates vertically. In a frame of reference attached to the tank, the gravity appears to oscillate which triggers a parametric instability which generates in turn internal gravity waves. Though the forcing mechanim does not impose any length scale, it eventually appears that the size of the tank provides the scale selection: only the largest wavelength was excited.

When Coriolis effects are important, the waves are named "inertia-gravity waves", with dispersion relation $\Omega^2 = N^2\cos^2\theta + f^2\sin^2\theta$ where f is twice the projection of the angular vector of the Earth onto the local vertical. In a laboratory experiment, f is

---

[2]By contrast, in the stratified shear flow problem for instance, the basic flow is made of a velocity profile of the form $U\tanh(z/\delta)$, which imposes both a velocity scale (U) and a length scale (δ), so that a time scale (δ/U) is imposed as well.

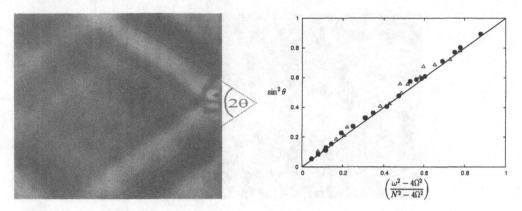

Figure 8. (a) Laboratory experiment of a vertically oscillating sphere in a stably-stratified rotating medium with constant N and f (half a domain is shown due to symmetry). (b) Verification of the dispersion relation (3.6) for different values of N and f. Note that θ denotes here the angle that the wave vector makes with the vertical and that the intrinsic wave frequency is denoted as ω (from Peacock and Tabaei, 2005).

twice the angular frequency of the rotating tank. It follows that Ω now becomes double-bounded: $f^2 \leq \Omega^2 \leq N^2$. The dispersion relation can also be written as

$$\cos^2\theta = \frac{\Omega^2 - f^2}{N^2 - f^2} \tag{3.6}$$

so that, again, one angle should be selected in a rotating stably-stratified medium by a forcing mechanism with frequency Ω. The laboratory experiment by Peacock and Tabaei (2005) nicely illustrates this relation (Figure 8).

3.4 The Kelvin-Helmholtz instability in a stably-stratified fluid

Evidence for the Kelvin-Helmholtz instability in natural media. The Kelvin-Helmholtz instability is also described by the linearized Boussinesq equations, when the basic flow is made of a velocity profile of the hyperbolic (or error function) type in a stably-stratified medium.

The Kelvin-Helmholtz instability develops at the interface between two fluids flowing horizontally at different velocities. This configuration is strongly unstable when the two fluids have the same density: a perturbation amplifies as soon as the Reynolds number associated with the shear flow is of order 1 or larger. As the perturbation amplifies, nonlinear effects come into play which make the instability saturate. The nonlinear development of the instability is manifested as Kelvin-Helmholtz vortices. Examples for this instability are ubiquitous in geophysical flows, whether in the atmosphere (Figure 9) or in the ocean (Figure 10). In the former case, the instability results from a vertically sheared wind; in latter case, the instability is due to tidal motion over a sill.

Figure 9. Kelvin-Helmholtz instability visualized by clouds on top of a mountain.

The Richardson number. A shear flow may exist in a constant N stably-stratified medium but a configuration most encountered in real flows is the one when the two fluids have also different densities (the lower fluid being of course heavier), leading to a vertical density profile of the hyperbolic tangent type, like the velocity profile. In this case, the density difference between the two fluids comes into play and the Froude number is no longer the most appropriate parameter to account for stratification effects. A bulk (or global or overall) Richardson number is rather introduced, whose definition comes naturally when the Boussinesq equations are made non-dimensional. The Richardson number is a measure of the stabilizing effect of buoyancy relative to the destabilizing effect of the shear.

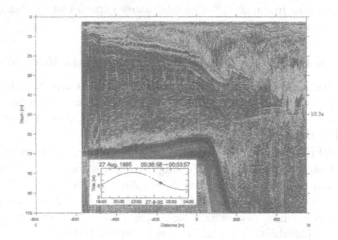

Figure 10. Kelvin-Helmholtz instability due to tidal motion over a sill (from Farmer and Armi, 1999).

Let us start from the Boussinesq equations linearized about a basic state consisting of the stratified shear flow $(U(z), R(z))$ (see the final set of equations at the end of section 2). We nondimensionalize these equations by choosing the following scales: U, the velocity scale, is half the velocity difference across the shear flow; δ, the length scale, is the thickness of the shear flow; $\Delta\rho$, the density scale, is half the density difference across the shear flow. T, the time scale, is equal to δ/U. It follows that the pressure scale is $\rho_0 U^2$.

The non-dimensional equations are (using the same notation for dimensional and dimensionless variables):

$$\frac{\partial u'}{\partial t} + U\frac{\partial u'}{\partial x} = -w'\frac{dU}{dz} - \frac{\partial p'}{\partial x} + \frac{1}{Re}\nabla^2 u' \tag{3.7}$$

$$\frac{\partial v'}{\partial t} + U\frac{\partial v'}{\partial x} = -\frac{\partial p'}{\partial y} + \frac{1}{Re}\nabla^2 v' \tag{3.8}$$

$$\frac{\partial w'}{\partial t} + U\frac{\partial w'}{\partial x} = -\frac{\partial p'}{\partial z} - J\,\rho' + \frac{1}{Re}\nabla^2 w' \tag{3.9}$$

$$\frac{\partial \rho'}{\partial t} + U\frac{\partial \rho'}{\partial x} = -\frac{dR}{dz}w' + \frac{1}{Pr\,Re}\nabla^2 \rho' \tag{3.10}$$

$$\nabla.\vec{u'} = 0. \tag{3.11}$$

These equations involve three non-dimensional parameters: the Reynolds number, $Re = \dfrac{U\delta}{\nu}$, the Prandtl number and the bulk Richardson number $J = g\dfrac{\Delta\rho}{\rho_0}\dfrac{L}{U^2}$.

The Richardson number criterion. When the two fluids have different densities such that the lower fluid is heavier, the instability cannot develop whatever the density difference: as fluid particles raise because of the instability, kinetic energy is converted into potential energy so that, intuitively, a too strong density difference prevents the instability from developing. Chandrashekar (1961) derived an heuristic criterion for local instability based upon energetic considerations, which was demonstrated rigorously by Howard (1961) (see also Drazin and Reid, 1981, p. 326). The energetic considerations compare kinetic and potential energies, that is nonlinear and stratification effects, which relates to the Richardson number. Since the bulk Richardson number does not provide any information about the local stability of the shear flow, a local Richardson number was introduced for this purpose:

$$Ri(z) = -\frac{g}{\rho_0}\frac{\frac{dR}{dz}}{\left(\frac{dU}{dz}\right)^2}. \tag{3.12}$$

The stability criterion is the following: a necessary condition for instability is $Ri(z) < 1/4$ somewhere in the fluid. It has been shown by Hazel (1972) that this condition is necessary and sufficient for a velocity and a density profile of the hyperbolic tangent type.

4 Instability of internal gravity waves in a constant N fluid at rest

4.1 Resonant interaction theory

Resonant interactions in the fluid dynamical context were discovered independently by Phillips (1960) and Hasselmann (1962) for surface gravity waves and progressively extended to internal gravity waves by considering, in particular, the interaction of two surface waves with an internal gravity wave (Ball, 1964); the case of an internal wave triad was eventually considered by Phillips (1966) and Hasselmann (1967). Extensive work was next performed in the oceanographic context (see the reviews by Phillips, 1981; Muller et al., 1986; Staquet and Sommeria, 2002). It was shown in particular that energy transfers in the spectrum of large scale, low frequency internal gravity waves in the ocean (the Garrett-Munk spectrum) mostly consist of resonant interactions (McComas and Bretherton, 1977). In the atmosphere, the decrease of density with altitude amplifies the waves, so that resonant interactions are only rarely expected (a significant fraction of the wave energy is still backscattered downward, see f.i. Barat and Cot, 1992). Perhaps one exception is the Antartic atmospheric boundary layer, which is surmonted by a very cold layer of air in which the waves may be trapped and interact resonantly (Rees et al., 2000).

We briefly summarize this theory in the following, which we illustrate by results from Koudella and Staquet (2005).

Wave steepness. In the following, we consider a monochromatic internal gravity wave $(\vec{k}, \Omega)$. We show below that fluid motions may be assumed to occur in the propagation plane $(\vec{k}, \vec{g})$ of the wave if the wave amplitude is infinitely small. Henceforth we assume two-dimensional fluid dynamics in the vertical (x, z) propagation plane. Let

$$(u, w) = (A_u, A_w) \, exp \, [\, I(\vec{k}.\vec{x} - \Omega t)\,] \quad + \quad c.c. \tag{4.1}$$

denote the velocity field induced by the wave, where $\Omega = F(\vec{k})$ obeys the dispersion relation (3.5).

The normalized amplitude, or steepness, of the wave is defined as:

$$s = \frac{u_{max}}{c_x}, \tag{4.2}$$

where c_x is the phase velocity in the x-direction. As illustrated in Figure 11, this definition implies that isopycnals (i.e. constant density surfaces) are nowhere overturned when $s < 1$ (the wave is also said statically stable) and that the isopycnals are locally overturned when $s > 1$ (statically unstable wave). The case $s = 1$ implies that vertical isopycnals have locally formed. Since $u_{max} = A_u$ and $c_x = \dfrac{\Omega}{k_x} = \dfrac{N}{|\vec{k}|}$, the steepness is also defined as

$s = \dfrac{A_u |\vec{k}|}{N}$ for a monochromatic wave. Hence, taking A_u as a velocity scale for the wave, $|\vec{k}|$ as a length scale and N^{-1} as a time scale, the steepness is simply the wave Froude

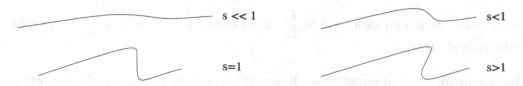

Figure 11. Illustration of the physical meaning of the steepness s of a monochromatic wave, defined by (4.2). A constant density line is drawn for different values of s.

number. Let us show that fluid motions may be assumed to occur in the propagation plane $(\vec{k}, \vec{g})$ of the wave for an infinitely small wave amplitude. In three dimensions, the equations of motion for the vorticity $\vec{\xi} = \nabla \times \vec{u}$ are expressed by:

$$\frac{\partial \vec{\xi}}{\partial t} + \vec{u} \cdot \nabla \vec{\xi} = \vec{\xi} \cdot \nabla \vec{u} - \nabla \times \frac{g}{\rho_0} \tilde{\rho} \vec{e}_z + \nu \nabla^2 \vec{\xi}. \tag{4.3}$$

The magnitude of the advective term $\vec{u} \cdot \nabla \vec{\xi}$ or of the vortex stretching/tilting term $\vec{\xi} \cdot \nabla \vec{u}$ relative to the magnitude of the baroclinic term $\nabla \times \frac{g}{\rho_0} \tilde{\rho} \vec{e}_z$ scales as the square of the wave steepness. This shows that, when $s^2 \ll 1$, the dynamics are not only weakly nonlinear but the fluid motions are mostly confined in a vertical plane if the vertical vorticity is initially zero. The linear stability analysis of a monochromatic internal gravity wave to three-dimensional perturbations by Klostermeyer (1991), later discussed in this paper, actually showed that this result holds as long as $s \lesssim 0.7$.

Resonance relations. The resonant interaction theory describes the interaction among a triad of internal gravity waves of infinitely small steepness ($s \ll 1$). Consider a triad of waves $(\vec{k}_0, \Omega_0)$, $(\vec{k}_1, \Omega_1)$ and $(\vec{k}_2, \Omega_2)$ with $\Omega_i = F(\vec{k}_i)$, $0 \leq i \leq 2$. Since the waves are nonlinearly interacting,

$$\vec{k}_0 + \vec{k}_1 + \vec{k}_2 = 0. \tag{4.4}$$

This relation is a spatial resonance relation. If, moreover,

$$\Omega_0 + \Omega_1 + \Omega_2 = 0, \tag{4.5}$$

with Ω_i of any sign, efficient interactions can occur among the wave triad. This relation is a temporal resonance relation. If the temporal resonance relation is not satisfied, the triad interaction produces harmonic perturbation, at frequency $\Omega_0 + \Omega_1 + \Omega_2$, without cumulative effects on wave amplitudes. These selective interactions are named 'resonant interactions'. They are selective because they imply both a spatial and a temporal resonance relation. It was shown by Hasselmann (1967) that "sum interactions" $|\Omega_1| + |\Omega_2| = |\Omega_0|$ are unstable. In the limit of small-scale secondary waves, that is $|\vec{k}_1|, |\vec{k}_2| \gg |\vec{k}_0|$, the instability is of the parametric type: $\Omega_1 \simeq \Omega_2 \simeq -\Omega_0/2$. Hasselmann also showed that "difference interactions" $|\Omega_1| - |\Omega_2| = |\Omega_0|$ are stable.

The energy and the quasi-momentum of the waves ($\vec{P}_i = \frac{E_i}{\Omega_i} \vec{k}_i$ for wave i) are conserved within the triad: $E_0 + E_1 + E_2 = $ constant and $\vec{P}_0 + \vec{P}_1 + \vec{P}_2 = $ constant. But the wave action, defined for each wave by $\frac{E_i}{\Omega_i}$, is not conserved: $\frac{E_0}{\Omega_0} + \frac{E_1}{\Omega_1} + \frac{E_2}{\Omega_2} \neq$ constant (Muller et al., 1986).

Resonant interaction equations. Resonant interactions among a triad occur over a slow time scale, inversely proportional to the wave steepness s (we recall that the theory is valid for $s \ll 1$). A slow time scale t_1 is therefore introduced, in addition to the (rapid) time scale t ($\simeq N^{-1}$) over which the waves evolve: $t_1 = \frac{t}{s}$. To describe the interaction, the amplitude and the phase of each wave is assumed to vary over the slow time scale t_1. A two-time scale asymptotic expansion of the velocity and density fields is performed and, at first order in s, the evolution equations of the amplitudes A_i are obtained (f.i. McEwan and Plumb, 1977):

$$\dot{A}_0 + T_0\, A_0 = S\, \Omega_0\, \overline{A}_1\, \overline{A}_2, \tag{4.6}$$

$$\dot{A}_1 + T_1\, A_1 = S\, \Omega_1\, \overline{A}_0\, \overline{A}_2, \tag{4.7}$$

$$\dot{A}_2 + T_2\, A_2 = S\, \Omega_2\, \overline{A}_0\, \overline{A}_1, \tag{4.8}$$

where T_i is the viscous damping rate, S is the interaction coefficient, which depends upon the wave vectors of the triad, and the bar denotes the complex conjugate.

4.2 Stability of a primary wave of infinitely small steepness

Theoretical results. The amplitude equations (4.6)-(4.8) can be used to investigate the stability of a monochromatic internal gravity wave, termed the primary wave. Let $(\vec{k}_0, \Omega_0)$ be the primary wave and assume that $A_1, A_2 \ll A_0$. Equations (4.6)-(4.8) may then be linearized about the basic state of the primary wave. An evolution equation for A_1 and A_2 is obtained (neglecting viscous effects for simplicity):

$$\ddot{A}_1 = S^2\, \Omega_1\Omega_2\, |A_0|^2 A_1. \tag{4.9}$$

It follows immediately that if $\Omega_1\Omega_2 > 0$, exponential growth of the secondary waves is possible, with growth rate equal to $S|A_0|\sqrt{\Omega_1\Omega_2}$. This condition was shown by Hasselmann (1967) to be equivalent to sum interactions, that is, the triad is unstable. The growth rate of the secondary waves is plotted as a function of $\vec{k}_1$ in Figure 12 for the run we consider below, which we shall refer to as run R1 for simplicity. For very large wavenumbers, resonance occurs through parametric instability and the growth rate saturates: there is no scale selection in the inviscid case, arbitrary small scales being excited by the instability. When viscous effects are taken into account, A_0 must be larger than a threshold amplitude for amplification to occur. The smallest scales are damped by viscosity, whatever its (finite) value, so that the viscosity provides the scale selection. In the case displayed in Figure 12, a wavenumber close to 5 is selected.

Numerical results. The theoretical results are illustrated by studying numerically the stability of a primary wave of large scale relative to the numerical domain (Koudella and Staquet, 2005). The Boussinesq equations are solved in a vertical plane for this purpose.

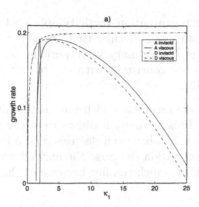

Figure 12. Theoretical predictions of the inviscid and viscous growth rates for an unstable triad as a function of wavenumber $|\vec{k}_1|$, for run R1 (from Koudella and Staquet, 2005).

Constant contours of the density and vorticity fields of the primary wave are displayed in Figure 13a. The primary wave vector is equal to $(1,1)$ in a $(2\pi)^2$ periodic domain and the wave steepness is equal to 0.36. Since a monochromatic wave is an exact solution of the inviscid Boussinesq equations, a small amplitude noise is superimposed upon the primary wave at initial time to trigger the instability. The Brunt-Väisälä frequency has a constant value equal to 1. This is run R1.

For simplicity, we assume that frequencies are all positive. Since we shall focus on unstable interactions, the temporal resonance relation should be written as $\Omega_0 = \Omega_1 + \Omega_2$ (sum interactions) and the associated spatial resonance condition is $\vec{k}_0 = \vec{k}_1 + \vec{k}_2$.

We found that a few triads are resonantly excited by the primary wave, among them being the triad $[\vec{k}_0 = (1,1), \vec{k}_1 = (3,7), \vec{k}_2 = (-2,-6)]$, which we refer to as triad T. It can be checked easily that this triad satisfies the spatial resonance condition. The temporal resonance condition is nearly satisfied: indeed $\Omega_0 = N\cos\theta_0 = 1/\sqrt{2} \simeq 0.707$, $\Omega_1 = N\cos\theta_1 = 3/\sqrt{58} \simeq 0.394$ and $\Omega_2 = -N\cos\theta_2 = 1/\sqrt{10} \simeq 0.316$, implying that $\Omega_0 - (\Omega_1 + \Omega_2) = -0.003$, which is smaller by two orders of magnitude than any of the three frequencies.

The latter values also show that the secondary waves $(\vec{k}_1, \Omega_1)$ and $(\vec{k}_2, \Omega_2)$ are resonantly excited by the primary wave through parametric instability. Indeed, one has $\Omega_1 \simeq \Omega_2 \simeq \Omega_0/2$. One would have Ω_1 much closer to Ω_2 if the wavenumbers of the secondary waves were larger but such small scale waves are damped by viscosity. The modulus of the secondary wave vectors in triad T is of order 5 instead, consistently with the scale selection observed in Figure 12.

How do these parametrically excited secondary waves manifest themselves in physical space? Constant contours of the perturbation vorticity are plotted at a given time in Figure 14a. (The dark band on the Figure is the region where the primary wave vorticity is negative and may ignored in this section.) We recall that the perturbation is made at initial time of the infinitely small amplitude noise, from where a few resonant triads emerge as time elapses. The perturbation vorticity appears to be organized as bands of alternate sign, which make an angle θ_1 such that $N\cos\theta_1 \simeq \omega_0/2$. Also, the sign of a given band changes sign after one primary wave period and recovers its sign after two periods. These observations are consistent with the perturbation being parametrically excited by the primary wave.

The fact that the bands are associated with parametrically excited secondary waves is confirmed in Figure 14b. This calculation differs from the one displayed in frame (a) by the value of the viscosity, which has been decreased by a factor 100. The same bands are visible, with the same inclination (because the excited waves have a frequency close to $\Omega_0/2$), but the wavelength is much smaller because of the smaller scale selection by the lower viscosity.

Energetics of the parametric instability.   In this section, we analyse the mechanism through which the perturbation grows. As usual in this type of study, we consider the energetics of the perturbation. The novel point here is that the spatio-temporal dependency of the forcing mechanisms of the perturbation is described (Koudella and Staquet, 2005). We start with a rotation of the reference frame (f.i. Mied, 1976) so that Z is aligned with the primary wave vector $\vec{k}_0$ in the new $(0, X, Z)$ frame. We decompose the velocity, density and pressure fields into a part associated with the primary wave and a part associated with the perturbation. The Boussinesq equations linearized about the primary wave state are, for the kinetic and potential energy of the perturbation (Lombard

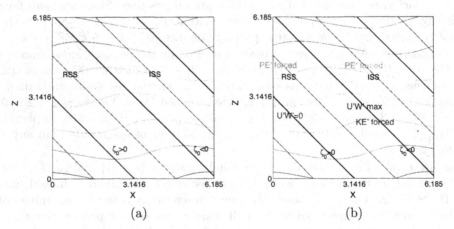

(a)   (b)

Figure 13. (a) Vorticity $\zeta_0 = dU_0/dZ$ and density fields of a monochromatic wave. Note the quadrature relationship between the two fields. A few isopycnals are shown. The dark (light) inclined bands are the regions where dU_0/dZ is negative (positive). (b) Same as (a), with regions where either the kinetic energy or the potential energy of the perturbation is forced. RSS (ISS): reduced (increased) static stability region (from Koudella and Staquet, 2005).

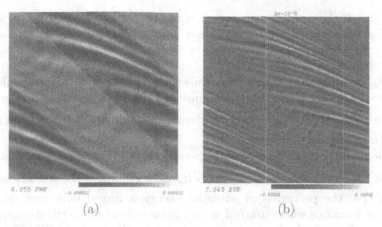

(a)   (b)

Figure 14. (a) Constant contours of the perturbation vorticity for run 1. (b) Same as (a) except that the viscosity has been decreased by a factor 100 (from Koudella and Staquet, 2005).

and Riley, 1996):

$$\frac{D}{Dt}\left[\frac{1}{2}\left(U'^2 + W'^2\right)\right] = -\nabla(\vec{U'}.P') - BF - U'W'\frac{dU_0}{dZ} \qquad (4.10)$$

$$\frac{D}{Dt}\left[\frac{1}{2}\frac{\rho'^2}{N^2}\right] = BF - \frac{1}{N^2}\rho'W'\frac{dR_0}{dZ}. \qquad (4.11)$$

where (U', W', P', ρ') are the perturbation fields in the rotated $(0, X, Z)$ frame and BF is the buoyancy flux. In this frame, the primary wave velocity is $(U_0, 0)$; we denote its density field by R_0.

These equations immediately show that the kinetic energy of the perturbation is forced by the term : $-U'W'\frac{dU_0}{dZ}$, which involves the shear of the primary wave (as for any unstable shear flow) and the potential energy is forced by the term: $\frac{1}{N^2}\rho'W'\frac{dR_0}{dZ}$, which involves the vertical gradient of the primary wave density field.

Let us consider the forcing of the kinetic energy. If we assume that the perturbation is made of a superposition of linear internal gravity waves, the correlation $U'W'$ is one-signed. It is not difficult to show here that $U'W' \geq 0$ in the rotated reference frame: the velocity field of the perturbation is aligned with the vorticity bands displayed in Figure 14 so that U' has the same sign as W'. On the other hand, dU_0/dZ oscillates in time and space. At a given position, this implies that the forcing term $-U'W'\frac{dU_0}{dZ}$ is positive, i.e. the perturbation is forced, when dU_0/dZ is negative; but the forcing would be negative when dU_0/dZ becomes positive at that position, implying that the perturbation would give back the acquired energy. How is a non zero net forcing possible? The point is that the perturbation will not give its kinetic energy back if the correlation term $U'W'$ is zero, or very small, during the half primary wave period where dU_0/dZ is positive. This will be the case if the perturbation energy is of potential form during this half-period. *It follows that optimum energy forcing occurs when the perturbation energy is of kinetic form during the half-period when $dU_0/dZ < 0$ and of potential form during the other half-period when $dU_0/dZ > 0$.*

Since dU_0/dZ is of the form $\exp(I\Phi_0)$, the energy of the perturbation should be of this form too. It follows that the velocity and density fields of the perturbation should behave as $(U', W', \sigma') \sim \exp(I\frac{\Phi_0}{2})$. In other words, the instability must be of the parametric type for optimal forcing to occur.

What about the potential energy of the perturbation? It can be shown that its forcing term $\frac{1}{N^2}\rho'W'\frac{dR_0}{dZ}$ is always positive. Since dR_0/dZ oscillates at the primary wave period, this implies that the perturbation potential energy is forced twice per primary wave period: at the locations where dR_0/dZ is maximum, where the vertical density gradient experienced by the perturbation is decreased, and at locations where dR_0/dZ is minimum, where the vertical density gradient experienced by the perturbation is increased. The former region is referred to as RSS (reduced static stability regions) and the latter as ISS (increased static stability regions). By contrast, as shown above, the kinetic energy is forced once, in regions where the primary wave shear dU_0/dZ is negative, and this is where vorticity bands are visible in Figure 14. Due to the continuous conversion of kinetic energy into potential energy, it can be shown that potential energy is converted into kinetic energy in a ISS region, the reverse occuring in a RSS region. As a consequence, the perturbation potential energy always increases in RSS regions.

A summary of the spatial occurence of these forcing mechanisms is displayed in Figure 13b.

As mentionned above, the perturbation growth is controlled by a few parametrically resonant triads, which may be assumed to be independent as they grow. Hence, the energetics of the perturbation may be reduced to that of a single resonant triad and approximate expressions of the perturbation energy and of energy transfer terms between the primary wave and the perturbation may be computed from this assumption. It can be shown in particular that the phase average over 2π of the kinetic energy transfer term $(-U'W'\frac{dU_0}{dZ})$ to the potential energy transfer term $(\frac{1}{N^2}\rho'W'\frac{dR_0}{dZ})$ is equal to $\cos(\theta_1 - \theta_0)$, that is, $\simeq 0.91$ for run R1. Hence, *potential energy is transferred at a higher rate than kinetic energy in an unstable primary wave of small amplitude.* This is actually a very general result in stratified turbulence, where it is found that potential energy transfer toward small scales is higher than kinetic energy transfers. The analogy can be pursued in the present case since parametric instability excites small scale waves. Two arguments can be provided to account for this behaviour : the incompressibility constraint, which applies to the velocity but not to the density field (Lesieur, 1997) and the subset of two-dimensional interactions which prevent kinetic energy from being transferred toward small scales (Holloway and Ramsden, 1988).

What does happen as the perturbation amplifies? The amplifying perturbation vorticity field in $dU_0/dz < 0$ regions may become unstable to a (large scale) Kelvin-Helmholtz instability. On the other hand, the perturbation potential energy continuously increases in RSS regions, eventually leading to unstable fluid layers; hence, a (small-scale) buoyancy-induced instability may also occur there. However, the unstable fluid layers are stabilized by the shear of the primary wave so that, in a two-dimensional computation, a Kelvin-Helmhotz instability eventually occurs. In a three-dimensional medium by contrast, the buoyancy-induced instability can develop in the plane perpendicular to the primary wave propagation plane because there is no shear there. Hence, in a three-dimensional fluid, both instabilities compete. Koudella and Staquet (2005) showed that the buoyancy-induced instability grows faster than the Kelvin-Helmholtz instability, unless viscous effects are strong enough, thereby confirming earlier heuristic analysis (Munk, 1981). It follows that *an unstable primary wave of small steepness eventually breaks down through a three-dimensional buoyancy-induced instability (if viscous effects are low enough)*.

4.3 Stability of a primary wave of finite steepness

Theoretical results. When the primary wave steepness is smaller than 1 but not infinitesimal, the use of a linear stability analysis against two-dimensional perturbations allows one to go beyond the resonant interaction theory. The early analyses neglected Coriolis effects. Drazin (1977) and Mied (1976) independently considered the case of a perfect fluid while Klostermeyer (1983) extended their results to the case of a real fluid. The fundamental conclusion of this two-dimensional analysis is that a primary wave is always unstable, whatever the associated Richardson number (that is whatever the intensity of stratification N relative to the wave-induced shear). The instability is again of the parametric type in the case of small scale perturbation. In the limit of a vanishing primary wave steepness, Drazin (1977) showed that the results of the linear stability analysis coincide with those of the resonant interaction theory.

By contrast, when the primary wave amplitude is so strong that isopycnals are overturned ($s > 1$), the analysis must be conducted in a three-dimensional framework. A linear stability analysis of this basic state to two- and three-dimensional perturbations was conducted by Klostermeyer (1991) ; only one orientation of the primary wave vector and of the plane of the three-dimensional perturbation was considered (this plane was perpendicular to the primary wave propagation plane $(\vec{k}, \vec{g})$). Klostermeyer (1991) found that the most amplified perturbation is three-dimensional and grows at small scale through a parametric instability mechanism. This pioneering work was complemented by Lombard and Riley (1996) and Sonmor and Klaassen (1996). Largest growth rates were found for perturbations at an angle oblique to the primary wave propagation plane but, as Lombard and Riley (1996) first realized, the features of the three-dimensional instability (wave number and amplitude) strongly depend upon the primary wave propagation angle (see their Figure 6).

Numerical results. These results are illustrated by three computations, referred to as runs R2, R3 and R4 (Koudella, 1999). The first one (run R2) is analogous to run R1 above, except that the initial primary wave now evolves in three dimensions and is perturbed by a three-dimensional noise of small amplitude (Figure 15a). Such a small steepness wave is unstable to parametric instability in its propagation plane implying that the flow remains two-dimensional during the instability growth (Figure 15b). The secondary resonant waves amplify most in RSS regions, as we showed it above. This leads to overturned isopycnals (s is locally greater than 1), which are unstable. A buoyancy-induced instability develops in the plane perpendicular to the propagation plane of the primary wave and the flow becomes three-dimensional (Figure 15c). Like the parametric instability, the buoyancy-induced instability amplifies arbitrary small scale waves so that the viscosity provides the scale selection mechanism. In Figure 15c, a wavelength equal to $\simeq 1/7$ of the domain size along the y-direction is visible. Because the instability has already reached a nonlinear stage in this frame, pairs of counter-rotating vortices (one pair per wavelength) are actually visible in the Figure. These small energetic scales are eventually dissipated by viscosity and the flow relaminarizes, because the primary wave is not forced.

It is remarkable that the validity of the resonant interaction theory extends much beyond its theoretical domain of validity, as we shall show it. We present in Figure 16 a run analogous to run R2 except that the initial primary wave steepness is twice larger ($s = 0.724$, run R3). An in run R2, the isopycnals locally amplify up to overturning while remaining two-dimensional: the primary wave becomes unstable through parametric instability during this stage (first two frames of the Figure). The overturned isopycnals become unstable through a buoyancy-induced instability (third frame). Because the viscosity is twice higher in this run than in run R2, a larger wavelength is selected along the y-direction, equal to $1/3$ of the domain size. The whole flow eventually becomes turbulent and dissipates.

Details of the formation of the buoyancy-induced instability in this run are presented in Figure 17. Snapshots of the density field are displayed in a plane permendicular to the propagation plane of the primary wave. Three wavelengths are thus visible in the second frame (instead of $\simeq 7$ for run R2). The nonlinear development of the instability gives rise to coherent structures with a mushroom form (fifth frame), which yield a pair of vortices of alternate sign. A zoom of the formation of the mushroom structures is displayed on the third row.

The formation of mushroom structures as a result of a buoyancy-induced instability has also been observed in laboratory experiments. Figure 18a stems from the experiment of Benielli and Sommeria (1998), already referred to in Section 3.3: it displays a visualization in a plane perpendicular to the propagation plane of a large scale internal gravity wave excited through parametric instability; two mushroom structures are visible. Figure 18b stems from the laboratory experiment of a stably-stratified shear layer performed by Schowalter et al. (1994).

The above results show that overturned isopycnals are spontaneously unstable to a three-dimensional buoyancy-induced instability. This is illustrated in Figure 19 through the temporal evolution of three isopycnal surfaces for an internal gravity wave of initial steepness larger than 1 (run R4). In this case, the viscosity is 50% larger than in run R3 and two wavelengths only appear only the y-direction. As previously noted, the whole wave field eventually breaks down through this instability. Such large amplitude internal gravity waves are encountered in the atmosphere, due to the rarefaction of air (lowering of density) as the wave propagates upward. Other mechanisms leading to a local amplification of the wave are discussed in the next section.

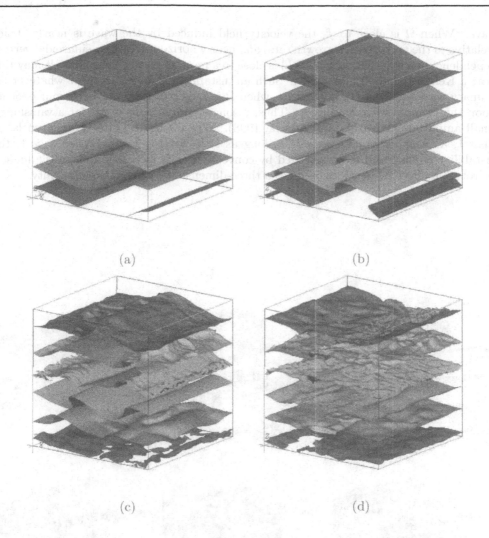

(a)   (b)

(c)   (d)

Figure 15. Breaking of an internal gravity wave with $s = 0.36$ (resolution 256^3). Six constant density surfaces are plotted at successive times: (a): $t = 16.1$ BVP, (b): $t = 38.2$ BVP, (c): $t = 42$ BVP, (d): $t = 44$ BVP. In frames (a) and (b), a two-dimensional secondary instability develops, leading to local overturnings of the isopycnals; in frame (c), a three-dimensional instability develops upon overturned isopycnals, and in (d), local breaking occurs. BVP is the Brunt-Väisälä period $2\pi/N$ (from Koudella, 1999).

Influence of Coriolis effects on the instability of a primary wave. When Coriolis effects are taken into account, the linear stability analysis of Dunkerton (1997) (see also references therein) and numerical experiments of Lelong and Dunkerton (1998) show that the instability mechanism depends upon the intrinsic frequency of the primary wave. When Ω is close to f, the velocity field induced by the wave is nearly steady (relative to the perturbation growth rate) and nearly horizontal (with a sinusoidal vertical dependency) so that this velocity field is close to a parallel steady shear flow: it may thus bear a Kelvin-Helmholtz instability. Such an instability is found to occur whatever the primary wave amplitude. By contrast, when Ω is larger than f, the behaviour depends upon the primary wave amplitude. Thus, resonant interactions occur for a vanishingly small amplitude (Miyazaki and Adachi, 1998). When s has a finite value, even larger than 1, a shear instability prevails as long as Ω is not too large. When $s > 1$, this instability is more and more modified by convection as Ω increases and, in the limit of no rotation effect ($\Omega \gg f$), it becomes a three-dimensional convective instability.

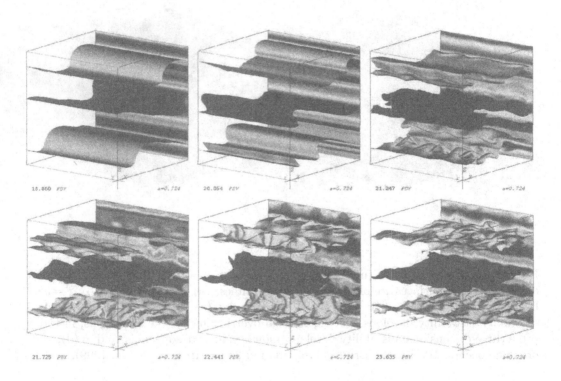

Figure 16. Breaking of an internal gravity wave with initial steepness twice larger than in Figure 15 (resolution 256^3). Three isopycnals are displayed at successive times. First row, from left to right, in unit of the Brunt-Väisälä period (BVP) $2\pi/N$: $t = 18.9$, 20, 21.25. Second row, from left to right: $t = 21.7$, 22.4, 23.6 (from Koudella, 1999).

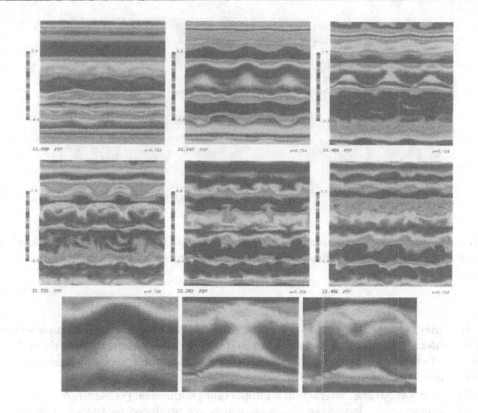

Figure 17. Vertical cross-sections of the density field for the primary wave displayed in Figure 16, at successive times. The last two frames are a zoom of the buoyancy induced instability. Visualizations are made in the plane perpendicular to the primary wave propagation plane (from Koudella, 1999).

5 Interaction of internal gravity waves with an ambient shear flow

5.1 Geophysical context

In geophysical flows, internal gravity waves generally interact with an ambient shear flow: for instance, the shear flow is the wind in the atmosphere (Bretherton, 1966) or a current in the ocean (Badulin et al., 1990). The former case is very common : the blowing of a wind over a mountain range generates lee waves which next interact with the wind. Recent research in the Sun has revealed that the differential rotation that exists in this star creates a very strong shear at the basis of the convective zone, which must interact with the waves that are produced there (Talon et al., 2002). How does this shear flow interact with the waves is unknown. An analogous interaction occurs when the waves encounter a temperature (or a density) front. Actually, when Coriolis effects

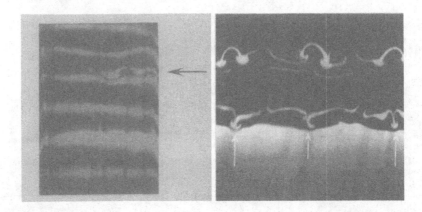

Figure 18. Laboratory experiments in which a buoyancy-induced instability develops (see text for details). Left frame: from Benielli and Sommeria (1998); right frame: from Schowalter et al. (1994).

come into play, such a front creates a vertical shear via the thermal wind balance so that the wave interact with both a density front and a shear. And when there is no ambient (large scale) shear or density front, as in the deep ocean, the wave field interacts with the shear flow it induces.

Why is the wave-vortex interaction an important geophysical problem? A very simple example is provided by lee waves. Since lee waves are produced by the wind, energy is extracted from the wind to generate these waves. As a result, the wind speed decreases over a mountain range because of this generation process, as if the mountain were exerting a drag force on the wind. This drag force has been introduced (as a subgrid-scale model) for the alpine range in the ECMWF[3] weather forecast model, thereby improving the weather predictions for this region (Lott, 1995). Another example may be provided by the permeability of the polar vortex. The polar vortex forms over the pole in the winter hemisphere and is a quasi-two-dimensional stratospheric vortex. It extends from the bottom of the stratosphere ($\simeq$ 8 kms at the pole) up to an altitude of 70 kms or so, while its horizontal extent is of a few thousand kms. Such a quasi-two-dimensional vortex is known to be a barrier to transport: laboratory experiments by Heijst et al. (1991) for instance show that dye injected in an isolated quasi-two-dimensional vortex remains capped inside the vortex and do not mix with the fluid outside the vortex. Atmospheric measurements at mid-latitudes still led to the suspiscion that air possibly leaks outside the polar vortex and a mechanism was seeked for to explain this leakage. McIntyre (1995) suggested that the interaction of inertia-gravity waves (which are internal gravity waves subjected to rotation) with the vortex edge could result in an irreversible transport of mass across the vortex edge. We finally report about a subtle mechanism of wave-vortex interactions in the ocean. Klein et al. (2003) showed that meso-scale eddies (which are

---

[3]European Center for Meteorological Weather Forecast.

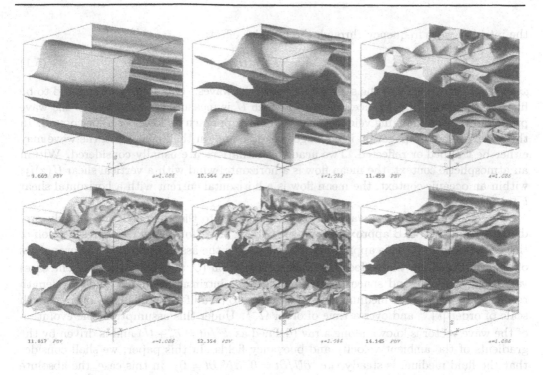

Figure 19. Breaking of an internal gravity wave with overturned isopycnals at $t = 0$ ($s = 1.086$, resolution 256^3). Three constant density surfaces are plotted at successive times. First row, from left to right, in unit of the Brunt-Väisälä period (BVP): $t = 9.7$, 10.6, 11.5. Second row, from left to right: $t = 11.8$, 12.3, 14.1 (from Koudella, 1999).

responsible for the transport of heat toward the poles), through their effects on near-inertial ($\Omega \simeq f$) oscillations, can induce a spatially heterogeneous diffusion that depends on the vorticity sign. In turn, the heterogeneous diffusion breaks the cyclone-anticyclone symmetry on short time scales (compared to the diffusive time scales) and favors the emergence of cyclonic structures.

5.2 Ray theory

In the following, we focus upon the interaction in an infinite medium of internal gravity waves with an ambient large scale horizontal shear flow $\vec{U}$, and, when rotation is present, with a density front as well.

We consider a monochromatic wave of intrinsic frequency Ω and wave vector $\vec{k}$ (or a wave packet with main intrinsic frequency and wave vector Ω and $\vec{k}$). The intrinsic frequency is the frequency measured in a frame of reference attached to the ambient flow. The simplest, and main, effect of the interaction on the internal wave is the change of the intrinsic wave frequency through the Doppler effect:

$$\omega_0 = \Omega + \vec{k}.\vec{U}, \tag{5.1}$$

ω_0 being the frequency of the source that emits the waves. This source is supposed to be fixed in the frame of reference relative to which $\vec{U}$ is measured. Ω changes as the wave propagates in the changing velocity field, and may approach its lower or upper bound. In this case, further propagation is no longer possible and, in the linear limit, the wave may either be trapped or reflected. Two academic situations are usually considered. Within an atmospheric context, the mean flow is a horizontal wind with a vertical shear $U(z)\vec{i}_x$; within an oceanic context, the mean flow is a horizontal current with a horizontal shear $U(y)\vec{i}_x$.

The change in Ω as the wave packet propagates into the current is most easily predicted within the WKB approximation (see Olbers, 1981, for a very clear presentation of this approximation). This approximation relies upon the assumption that the properties of the fluid medium that affect the wave propagation ($\vec{U}$ and N, in the present case) vary slowly in time and space relative to the wave intrinsic frequency and wavelength respectively. Hence, the medium may be assumed to be uniform and steady over a length scale of order $|\vec{k}|^{-1}$ and over a time of order Ω^{-1}. Under this assumption, the evolution of the wave vector is known along a ray (defined as $d\vec{x}/dt = \vec{c}_g + \vec{U}$) and is driven by the gradients of the ambient velocity and buoyancy fields. In this paper, we shall consider that the fluid medium is steady (*i.e.* $\partial U/\partial t = 0$, $\partial N/\partial t = 0$). In this case, the absolute frequency is constant along a ray: $d\omega_0/dt = 0$, where $d/dt = \partial/\partial t + (\vec{c}_g + \vec{U}).\nabla$ denotes the material derivative following a ray. The equations governing the refraction of the wave vector along a ray, known as the ray equations, are:

$$\frac{dk_i}{dt} = -\frac{\partial\Omega}{\partial N}\frac{\partial N}{\partial x_i} - k_j\frac{\partial U_j}{\partial x_i}, \tag{5.2}$$

that is, in the present case, with $\mathbf{U} = (U(y,z),0,0)$ and $N(y,z)$

$$\frac{dk_x}{dt} = 0 \tag{5.3}$$

$$\frac{dk_y}{dt} = -\frac{\partial\Omega}{\partial N}\frac{\partial N}{\partial y} - k_x\frac{\partial U}{\partial y}, \tag{5.4}$$

$$\frac{dk_z}{dt} = -\frac{\partial\Omega}{\partial N}\frac{\partial N}{\partial z} - k_x\frac{\partial U}{\partial z}. \tag{5.5}$$

As for the relative frequency, it changes along a ray according to the equation

$$\frac{d\Omega}{dt} = -c_{gi}k_x\frac{dU}{dx_i}, \tag{5.6}$$

where c_{gi} is the component of the group velocity along the direction of inhomogeneity x_i.

Changes of the wave amplitude are inferred from the conservation of wave action. For any slowly varying background, the action $A = E/\Omega$, where E is the wave energy,

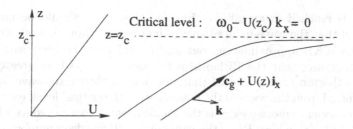

Figure 20. Sketch of a wave packet approaching a critical level $z = z_c$, within the WKB approximation. Two rays are shown in the vertical (x, z) plane, with $\vec{c}_g + U(z)\vec{i}_x$ being the absolute group velocity. The intrinsic frequency of the wave packet, $\omega_0 - U(z_c)k_x$, from the Doppler relation, vanishes at the critical level.

satisfies the conservation equation (Bretherton, 1968):

$$\frac{\partial A}{\partial t} + \nabla.[(\vec{c}_g + \vec{U})A] = 0. \tag{5.7}$$

Equation (5.7) implies that the action contained in a small volume δV moving with the absolute group velocity is conserved, that is

$$\frac{d(A\delta V)}{dt} = 0. \tag{5.8}$$

The form of the WKB theory we use is the approximation of geometrical optics but, for simplicity, the terminology *WKB approximation* will be employed hereafter.

5.3   Horizontal mean flow with a vertical shear $U(z)\vec{i}_x$

This situation was first investigated without rotation effect, when a wave packet propagates upwards in a vertical shear flow such that its intrinsic frequency decreases. If there exists an altitude at which Ω vanishes, the wave cannot propagate beyond this altitude, known as the critical level. Hence the critical level acts as a wave filter. This level is also defined as the altitude at which the component of the phase velocity of the wave along the wind direction equals the wind velocity (Figure 20).

The first theoretical approach to this problem was performed by Bretherton (1966), who considered the Boussinesq equations in a non-rotating frame, linearized about a basic state defined by the wind and a linear stratification, under the WKB approximation. Note that this approximation implies that the Richardson number $(Ri = N^2/(dU/dz)^2)$ is much larger than 1 everywhere (Olbers, 1981). For a wave packet approaching the critical level, Bretherton (1966) showed that the vertical component of the group velocity c_{gz} decreases as η^2, where η is the distance of the wave packet to the critical level, while k_z increases as η^{-1}. From the dispersion relation (3.5), $\Omega \to 0$ (since k_x and k_y are constant). Even if these results are correct as the critical level is approched, the WKB approximation diverges in the immediate vicinity of that level. The theory predicts that this level is reached in an infinite time (so that there should be no transmitted component) and that the wave-induced energy increases without bound, as η^{-1}.

Relaxing the WKB approximation, but still considering the linearized equations of motions and assuming that the Richardson number is everywhere greater than 0.25, Booker and Bretherton (1967) showed that, for a monochromatic wave, momentum is actually transferred from the wave to the mean flow at the critical level, except for a weak transmitted component whose energy is the incident wave energy reduced by the factor $\exp[-2\pi(Ri_c - 1/4)^{1/2}]$, where Ri_c is the value of the Richardson number at the critical level. The transfer of momentum to the mean flow is manifested as a discontinuity in the vertical flux of wave-induced horizontal momentum $\overline{\rho u'w'}(z)$ at the critical level (where the bar denotes an average over a horizontal wavelength); this flux is otherwise constant with altitude in the absence of critical level (Eliassen and Palm, 1961). No component is reflected in this linear limit.

For a high enough wave amplitude, the incoming wave-induced energy is higher than that absorbed by the shear flow and, consequently, the energy density increases in the neighbourhood of the trapping level. Breaking eventually occurs, thereby partly dissipating the accumulated energy (since momentum is conserved however, momentum will be locally deposited by the breaking waves). This process is well-known to occur in the atmosphere. However, as noted by McIntyre (2000), there is a subtle effect which implies that, unless viscous effects are too strong, the linear theory always fails after some time: as the wave approaches the critical level, its intrinsic phase velocity decreases faster than the wave-induced velocity along the wind direction does because of absorption so that their ratio, which is one measure of the wave steepness, increases. Hence, breaking may occur as well.

This situation was studied in detail through several experimental works (f.i. Koop, 1981; Koop and McGee, 1986) and is exemplified in Figure 21a for a weak amplitude wave and in Figure 21b for a large amplitude wave. Koop and McGee (1986) found that when breaking occurs, it does so through a two-dimensional Kelvin-Helmholtz instability when no critical level exists and through a three-dimensional buoyancy-induced instability when there is a critical level. The former behaviour can be explained as follows: as the waves propagate in the shear flow, the intrinsic frequency decreases so that wave-induced motions are close to horizontal, slow and of small steepness; such motions are therefore prone to Kelvin-Helmholtz instability. The latter behaviour is consistent with the general analysis of Lelong and Dunkerton (1998), who show that a gravity wave of low frequency and high steepness breaks through a buoyancy-induced instability. In numerical experiments, the breaking problem was first addressed in a two-dimensional vertical plane. The idea was that, as $k_z \to 0$, a strong wave-induced vertical shear develops (which actually dominates the mean flow shear), so that destabilisation should occur via a Kelvin-Helmholtz instability. The two-dimensional simulations realized by Winters and D'Asaro (1989) actually revealed an unexpected behaviour once the strong wave shear has built up, consisting of a long regime of overturned isopycnals without breaking. This led Winters and Riley (1992) to design a model for a gravity wave packet in the neighbourhood of the critical level, as a basic state for a linear stability analysis. They found that the most unstable modes are both of the Kelvin-Helmholtz type (occuring in the propagation plane of the wave packet) and of the convective type (occuring in a

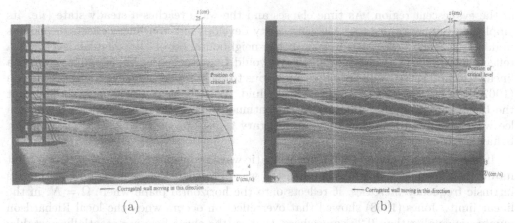

(a)　　　　　　　　　　　　　　　　(b)

Figure 21. Laboratory experiment of an internal gravity wave interacting with a vertical shear flow. Both the shear flow and the wave field are created by a moving topography. (a) The wave amplitude is small enough for complete absorption of the wave field by the shear flow to occur; (b) for a larger wave amplitude, the wave breaks below the critical level (from Koop and McGee, 1986).

plane perpendicular to it). Three-dimensional numerical simulations (Fritts et al., 1994; Winters and D'Asaro, 1994) later confirmed that the dominant instability is a three-dimensional convective instability, which the two-dimensional simulations were unable to reproduce. Only Winters and D'Asaro (1994) studied the influence of the breaking of a wave packet upon the fluid medium, for $Ri = 25$ and for three values of the wave packet initial amplitude. The main result is as follows: the energy carried by the wave packet is first (in time) transferred to the mean flow, the more so when the initial wave amplitude is smaller, in agreement with linear theory. A reflected component is next produced, whose energy content grows with the initial wave amplitude (no component seems to be transmitted, whatever the amplitude). Consequently, as noted by Winters and D'Asaro (1994), very little energy is left for fluid mixing and for potential vorticity changes.

The work of Booker and Bretherton (1967) was extended to rotating flows by Jones (1967) and by Wurtele et al. (1996). Using a linearized approach, Jones (1967) showed that, in addition to the classical critical level (characterized by $\Omega = 0$), two additional singular levels exist, corresponding to $\Omega = \pm f$. His paper contains another important result: in a rotating fluid, the vertical flux of wave-induced horizontal momentum is no longer conserved (away from the critical level). It should be replaced by the vertical flux of wave-induced angular momentum for this conservation property to hold again (away from the singular levels). Wurtele et al. (1996) further showed that the wave becomes evanescent in the neighbourhood of the critical level (since the wave can only propagate for $\Omega^2 > f^2$) so that the only effective singular levels are those where $\Omega^2 = f^2$. Wurtele et al. (1996) also investigated numerically the situation of a time developing wave, being emitted from a source at $t = 0$. In this case, the singularity develops in time as well : at early times, the propagating wave crosses the $\Omega = f$ (for instance) level and decays in the evanescent region. As time elapses and the wave reaches a steady state (*i.e.* its amplitude becomes constant), the singularity develops but non linear effects develop as well. As a result, the wave breaks in the neighbourhood of the singular level and a reflected component is emitted; the wave would be absorbed at the singular level in a linear regime. These results are thus analogous to those found by Booker and Bretherton (1967) at a critical level in a non rotating fluid. Wurtele et al. (1996) also showed that the behaviour is notably different for a continuous spectrum of frequencies (such as a lee wave generated by a flow over an arbitrary topography), in which case no singular behaviour is encountered.

In the discussion above, the monochromatic wave propagates in the shear flow so that its intrinsic frequency *decreases*. When the wave propagates in a shear flow so that its intrinsic frequency *increases*, it reflects onto the horizontal plane where $\Omega = N$, in the linear limit. Jones (1968) showed that overreflection occurs when the local Richardson number is smaller than 0.25 somewhere, that is, the shear flow is potentially unstable; the wave extracts energy from that flow when reflecting. When nonlinear effects come into play (due to the high initial amplitude of the wave for instance) and the background shear flow is stable, Sutherland (2001) showed that a horizontally periodic wave packet permanently deposits momentum to the mean flow at altitudes close to and below the reflecting level; when the wave packet is horizontally compact, a substantial part of the finite amplitude wave packet energy may be transmitted.

5.4 Horizontal mean flow with a horizontal shear $U(y)\vec{i}_x$

Theoretical results. The behaviour of an internal gravity wave packet in a horizontally sheared current $U(y)\vec{i}_x$ was studied thirty years ago within an oceanographic context, using the linear theory in the WKB approximation (Ivanov and Morozov, 1974; Olbers, 1981; Basovich and Tsimring, 1984; Badulin et al., 1985; Badulin and Shrira, 1993). Coriolis effects are ignored in all studies except in Olbers' work. When the wave packet enters into the current and propagates against it so that its intrinsic frequency increases (Eq. (5.6) with $x_i = y$), it cannot propagate beyond the position y_t at which $\Omega = N$. The mean flow being barotropic, this position actually is a vertical plane if N is constant; this plane is hereafter referred to as the trapping plane (Figure 22).

Since the properties of the medium in which the wave propagates vary only with y, k_x and k_z remain unchanged (as well as ω_0 as already noted, because the medium is steady). From the linear dispersion relation (3.5), one easily infers that k_y goes to infinity as $\Omega \to N$. More precisely, it can be shown from WKB theory that $k_y \sim \eta^{-1/2}$ as $\eta \to 0$, where η refers again to the distance of the wave packet to the trapping plane; moreover, $c_{gy} \sim \eta^{3/2}$ and c_{gx}, $c_{gz} \to 0$ as well (Staquet and Huerre, 2002). The latter property implies that the wave packet slows down in the neighbourhood of the trapping plane so that its energy density locally increases (in other words, the wave-induced energy accumulates in the neighbourhood of this plane). However, the WKB theory also predicts that this energy tends to infinity ($E \sim \eta^{-3/2}$) and that the trapping plane is reached in an infinite time. As in the critical level situation, the two latter results are unphysical and stem from the failure of the WKB theory in the immediate vicinity of the trapping plane. Olbers (1981) actually noted that, if $E \sim \eta^{-\mu}$, the asymptotic behaviour of the wave is

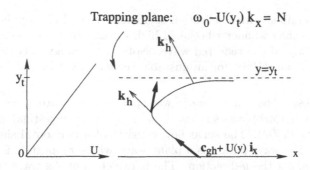

Figure 22. Sketch of a wave packet approaching a trapping plane, within the WKB approximation, and propagating against the current ($\vec{k}.\vec{U} = k_x.U < 0$). Two rays are shown in the horizontal (x, y) plane, with $\vec{c}_{gh} + U(y)\vec{i}_x$ being the component of the absolute group velocity in that plane. The intrinsic frequency of the wave packet, $\omega_0 - U(y_t)k_x$, from the Doppler relation, is equal to the Brünt-Väisälä frequency at the trapping plane (adapted from Staquet and Huerre, 2002).

regular or singular depending upon whether $\mu < 1$ or $\mu \geq 1$. In the former case, the wave reaches the plane in a finite time, and this plane is for instance a reflexion plane. In the latter case, the wave reaches the plane in an infinite time; very strong gradients of the wave-induced Reynolds stress form in the close neighbourhood of the plane, which yield momentum exchange with the mean flow. Information upon the actual wave behaviour when $\mu \geq 1$ can be obtained by solving the linearized equations of motions. Ivanov and Morozov (1974) thus found that the total wave-induced energy may indeed increase, as opposed to the critical level situation studied by Booker and Bretherton (1967). In this situation, *momentum is transferred from the shear flow to the wave*, so that the potential for wave breaking exists. The three-dimensional numerical study of Staquet and Huerre (2002), whereof a summary is provided below, shows that an inertia-gravity wave packet may indeed break in the neighbourhood of a trapping plane.

When the wave propagates along the current such that its intrinsic frequency decreases, k_y decreases as well. If k_y decays down to zero, Ω reaches a minimum value Ω_{min}, obtained by setting k_y to 0 in the dispersion relation. The behaviour of the wave as $\Omega \to \Omega_{min}$ may be guessed by using WKB theory. Note that the theory becomes less and less valid as the plane gets closer since the wavelength along the y-direction increases. Assuming that the theory remains valid, $k_y \sim \eta^{1/2}$, $c_{gy} \sim \eta^{1/2}$ and $E \sim \eta^{-1/2}$ as $\eta \to 0$. Here, according to Olbers' (1981) argument, $\mu = 1/2$ so that the $\Omega = \Omega_{min}$ plane is a reflexion plane for the wave.

Ollers et al. (2003) solved numerically the equation for the amplitude of a hydrostatic internal gravity wave emitted away from the shear flow and propagating toward it. The wave behaviour (transmitted, reflected, over-reflected) depends upon the stability of the shear flow. When the latter flow is inertially stable, the wave is always reflected, with a possible transmitted component. When the shear flow is inertially unstable by contrast, over-reflection is possible, with a reflection coefficient up to 3.25 (for the case considered in the paper). One may wonder whether a high enough reflexion coefficient would not lead to the instability of the reflected wave. Note the analogy between this behaviour and that found by Jones (1968) for an unstable vertical shear flow.

Numerical results. The behaviour of a wave packet in a rotating, constant-N medium propagating into a barotropic shear flow $U(y)\vec{i}_x$ has been investigated numerically by Staquet and Huerre (2002). The shear flow consists of a horizontal shear layer (with a tanh profile) while the wave packet is a plane wave whose amplitude is modulated by a gaussian function along the y-direction. The parameters of the wave (its wave vector), of the shear flow (its maximum amplitude) and of the medium (the Brunt-Väisälä and Coriolis frequencies) are chosen so that (i) the wave intrinsic frequency increases as the wave propagates in the shear flow and (ii) a trapping plane exists.

The wave behaviour is illustrated in Figure 23 for a cyclonic shear flow (that is, the vorticity of the shear flow is of the same sign as the Coriolis frequency). Constant contours of the density field are displayed at successive times in a vertical (y, z) plane. At $t = 0$ (Figure 23a), the wave packet is hardly visible due to its small steepness ($s = 0.26$). Because the horizontal shear flow does not displace the isopycnals, it is not visible either. Since N is constant, the trapping plane is a vertical plane and its intersection with the (y, z) plane is marked with a vertical line in the Figure. The wave packet exhibits two major changes as it propagates toward the shear flow: the isopycnals steepen and the wave amplitude increases. The former effect is accounted for by noting that, as k_y increases, the incompressibility condition reduces to $k_y u_y + k_z u_z \simeq 0$. Hence, the vector (k_y, k_z) is perpendicular to the phase lines in the (y, z) plane. Since $k_y \to +\infty$ while k_z remains constant, the phase lines steepen. The second effect results from the trapping of the wave. The local increase of the wave amplitude makes the wave packet break (Figure 23e) and small-scale motions are produced. The latter motions are quickly dissipated however because the primary wave packet is not forced. As well, the shear flow is hardly modified by the momentum deposit that occurs during wave breaking.

The stage of the flow that follows breaking dramatically changes when the shear flow is anticyclonic (Figure 24). Indeed, in this situation, the shear flow is subjected to an inertial instability, through which small-scale motions are most amplified. The point is that the small-scale motions resulting from wave breaking act as a perturbation to the shear flow, which triggers the instability. The medium is therefore considerably modified by the breaking of the wave in this situation, because it initiates the inertial instability of the (very energetic) shear flow. The latter instability results in momentum and mass transport: the shear of the background flow is weakened and a passive scalar is transported across the trapping plane, namely across the shear flow.

5.5   Horizontal mean flow with both a horizontal and a vertical shear

What does happen if a wave packet in a stably-stratified rotating medium interacts with a (thermal wind) balanced shear flow $\vec{U}(y, z)$ involving both a horizontal and a vertical shear? As discussed above, the situation is not simple. For instance, when the intrinsic frequency Ω increases and approaches N, the wave packet should be reflected

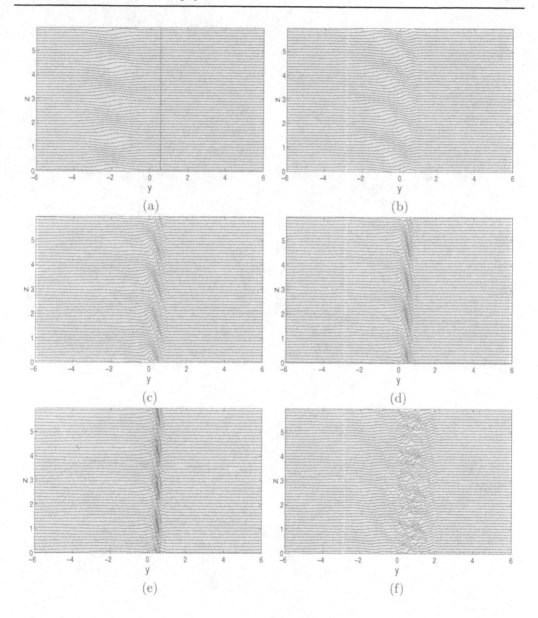

Figure 23. Interaction of a wave packet (confined along the y-direction only) with a barotropic cyclonic shear layer. Constant contours of the density field are displayed at successive times, expressed in Brunt-Väisälä periods; (a) $t = 0$, (b) $t = 1.9$, (c) $t = 3.8$, (d) $t = 4.6$, (e) $t = 5.6$, (f) $t = 11.4$. The vertical line in frame (a) marks the trapping plane.

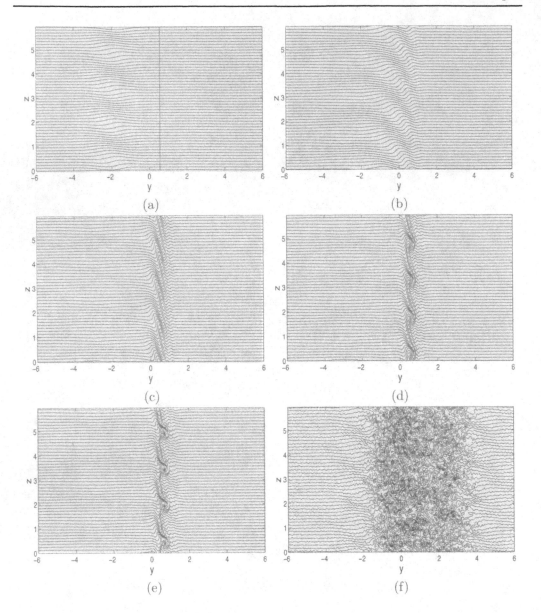

Figure 24. Interaction of a wave packet (confined along the y-direction only) with a barotropic anticyclonic shear layer. Constant contours of the density field are displayed at successive times, expressed in Brunt-Väisälä periods (a) $t = 0$, (b) $t = 1.9$, (c) $t = 3.8$, (d) $t = 4.6$, (e) $t = 4.8$, (f) $t = 9.4$. The vertical line in frame (a) marks the trapping plane (from Staquet and Huerre, 2002).

by the vertical shear $\partial U/\partial z$ but trapped by the horizontal shear $\partial U/\partial y$. Also, since the shear flow satisfies the thermal wind balance, a buoyancy field $B(y,z)$ exists as well such that $f\dfrac{\partial U}{\partial z} = -\dfrac{\partial B}{\partial y}$.

We have investigated this situation in a simple context, when a wave packet confined both in the y and z directions interacts with a horizontal shear layer with a sinusoidal vertical dependency (Edwards and Staquet, 2005). We chose the parameters such that Ω increases because of the horizontal shear of the background flow. We first explored the parameter range by solving the ray equations and performed three-dimensional direct numerical simulations (DNS) to investigate the influence of nonlinear effects on the behaviour of the wave packet.

A general behaviour is observed in the WKB theory, which is illustrated in Figure 25. The ray trajectories are displayed in frame (a) and the intrinsic frequency Ω is plotted versus time in frame (b). In frame (a), the wave packet at initial time is represented by a set of twenty rays aligned along two perpendicular segments. The $\Omega = N$ surface, plotted with a thick dashed line, displays two important locations when the interaction with the wave packet is considered: (i) where the local radius of curvature is minimum, corresponding to a maximum value of $\partial U/\partial y$ (location 1) and where the local radius of curvature is maximum, corresponding to a minimum value of $\partial U/\partial y$ (location 2). In the former case, the $\Omega = N$ surface is nearly vertical and is a trapping surface; in the latter case, the surface is nearly horizontal and is reflecting. All rays propagate toward the $\Omega = N$ surface and reach it, either in the neighbourhood of location 1 or of location 2. In the former case (location 1), the rays are trapped at the surface and propagate along it downward, with a nearly vertical group velocity, toward location 2. Note that the group velocity has strongly decreased when location 1 is reached. All rays sooner or later reach location 2 and reflect there. The rays then propagate in the interior of the shear flow within a wave guide made by the $\Omega = N$ surface. Frame (a) also displays grey points, at which the steepness of the wave packet exceeds 1. This suggests that breaking may occur there, resulting in irreversible mass and momentum transport. Note however that WKB theory is no longer valid at the trapping plane (and at a reflecting surface) and that, most importantly as we shall see, molecular effects have been ignored in this analysis.

DNS results are displayed in Figure 26 through constant contours of the fluctuating buoyancy field b'. It should be stressed that the steepness of the wave packet is twice smaller than in the baroclinic cases and that the varying horizontal shear $(\partial U/\partial y)$ is not larger. Hence, the wave-shear interaction (which scales like $s^2\partial U/\partial y$) is smaller by a factor 4 at least. The interaction between the shear flow and the wave packet is therefore weak and one may consider that the buoyancy fluctuations displayed in Figure 26 solely belong to the wave field. The DNS behaviour is close to the WKB prediction up to the time the wave packet reaches the trapping surface at location 2 (at $t \simeq 176 \simeq 28$ BVP). This is attested in frame b) where the intrinsic frequency predicted by WKB theory and the numerical simulation are compared. Molecular effects deeply change the subsequent wave packet behaviour in the DNS, for two reasons. First, k_y increases as the packet approches the trapping surface, implying that small scales along the y-direction are produced. These small scales are of course very sensitive to molecular effects. The

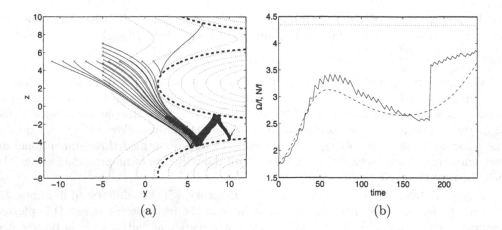

(a)   (b)

Figure 25. WKB predictions. (a) Trajectories of rays at $t = 127.3$ BVP (Brunt-Väisälä period) in a vertical (y, z) plane. The rays start from the initial wave packet location. The rays start from the initial wave packet location. The thick dashed line marks the intersection of the $\Omega = N$ trapping surface with the (y, z) plane and dotted lines represent contours of the shear flow velocity $U(y, z)$. Light grey circles are plotted at each time the steepness along a given ray exceeds the value of 1. (b) Temporal evolution of the intrinsic frequency Ω of the wave packet normalized by the Coriolis frequency f. Dashed line: WKB prediction for a central ray; solid line: DNS result at the packet centre; dotted line: N/f (upper bound of Ω/N). Results are compared over the duration of the DNS (t=240, that is 38.2 BVP) (from Edwards and Staquet, 2005).

second reason, connected to the first, is that the packet slows down as it approaches the trapping plane, which makes it also prone to molecular effects. As a consequence, the nearly steady, small-scale packet is dissipated locally and does not penetrate into the wave guide.

The general conclusion that can be drawn from this study is that a single wave packet is unlikely to modify its environment. In geophysical flows however, waves are most often generated by a permanent source (like the interaction of the tide with the topography in the ocean) or, at least, are emitted during a long time with respect to their intrinsic period. Since, in our study, the wave packet slows down as it approches the trapping surface, a continuous emission of such packets would result in their superposition in the neighbourhood of the reflecting surface, and possibly in breaking. Higher resolution simulations should also be conducted to reduce the influence of molecular effects on the wave packet behaviour.

6 Interaction of internal gravity waves with a boundary

6.1 Focusing by wall reflexion

In this section, we show that internal gravity waves can locally amplify by a simple geometrical effect first discovered by Phillips (1966), which may result in breaking and

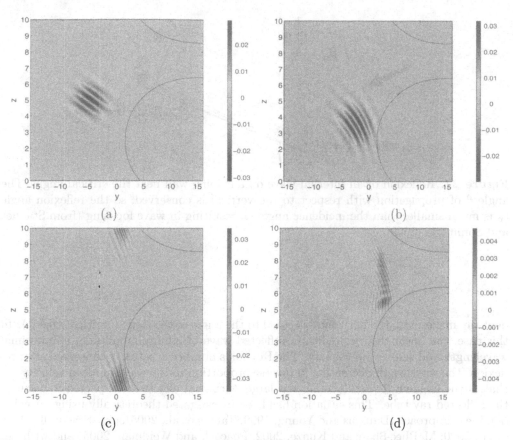

Figure 26. DNS results. Constant contours of the fluctuating buoyancy field b' are plotted in a vertical (y, z) plane at successive times (in unit of Brunt-Väisälä period): (a) t=0; (b) t=8.9, (c) t=13.4, (d) t=28. In frame (a), dotted lines represent contours of the shear flow velocity $U(y, z)$. The surface $\Omega = N$ is displayed with a solid line on all frames (from Edwards and Staquet, 2005).

turbulence. Coriolis effects are ignored in this paragraph.

When an internal gravity wave reflects onto a sloping boundary, the frequency is conserved upon reflection, as for any reflecting wave in a steady medium. The dispersion relation (3.5) thus implies a peculiar property (Phillips, 1966): the angle that the wave vector makes *with the horizontal* is preserved as well upon reflexion[4]. This geometrical property implies that a ray tube (made of curves everywhere tangent to the group velocity) reflecting on a sloping boundary may be focused (see Figure 27). Since the flux of energy across the tube section is preserved during reflexion, the energy flux increases after reflexion. A particularly interesting situation occurs when the angle that the inci-

---

[4]The Snell-Descartes law valid for non dispersive waves is recovered only when the slope is horizontal or vertical.

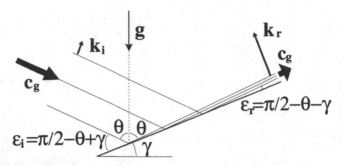

Figure 27. Reflexion of an internal wave on a sloping wall near the critical angle. The angle θ of propagation with respect to the vertical is conserved, so the reflexion angle ϵ_r is much smaller than the incidence angle ϵ_i, resulting in wave focusing (from Staquet and Sommeria, 2002).

dent ray makes with the horizontal is equal to the angle of the slope ("critical angle"). In this case, the linear theory predicts a reflected wave of infinite amplitude, infinitesimal wavelength and zero group velocity. The Reynolds number is actually conserved after reflexion (the wavelength decreasing in inverse proportion to the wave-induced velocity) so that, if the incident Reynolds number is large, turbulent effects are expected to occur in the reflected ray tube. This situation has been investigated theoretically using a weakly nonlinear approach (Dauxois and Young, 1999; Tabaei et al., 2005), experimentally (Ivey et al., 2000; McPhee-Shaw and Kunze, 2002; Peacock and Weidman, 2005) and with numerical simulations (Slinn and Riley, 1998; Zikanov and Slinn, 2001). Dauxois and Young (1999) predicted the generation of a harmonic component, which was recently confirmed by the laboratory experiment of Peacock and Weidman (2005), and the formation of an array of counter-rotating vortices along the wall. Slinn and Riley (1998) showed that a strongly turbulent boundary layer can form along the wall with efficient mixing property and, according to the laboratory experiment of McPhee-Shaw and Kunze (2002), this results in irreversible mass transport along the isopycnals and therefore away from the boundary, in so-called nepheloid layers (Figure 28) (see Staquet and Sommeria, 2002, for further discussion).

In a closed domain in which one wall (at least) is inclined to the vertical, the solutions of the linear equation are not longer regular solutions (seiches) but have a fractal structure. In this situation, the incident ray tubes focus at each reflexion on the sloping wall. It has been shown by Maas and Lam (1995) that the focused tubes actually converge toward an "attractor", namely toward a localized region in space (Figure 29a). A laboratory experiment has been performed to check the validity of these theoretical ideas (Maas et al., 1997): the occurence of the attractor is clearly visible in Figure 29b. The occurence of turbulence and mixing along the attractor is still an open question.

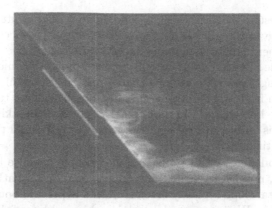

Figure 28. Laboratory experiments of an internal gravity wave reflecting on an inclined wall. Fluoresceine dye at the wall is transported along isopycnals, away from the wall, as a result of the breaking of the wave (from McPhee-Shaw and Kunze, 2002).

6.2   The internal tide

The internal tide is the field of internal gravity waves generated in the ocean by the interaction of the barotropic tide (the common tide) with the bathymetry. Munk and

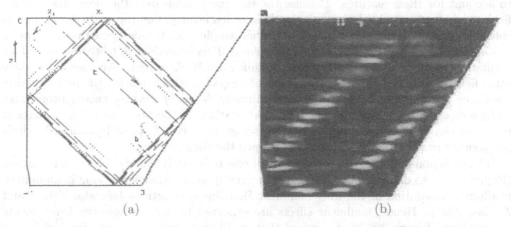

(a)   (b)

Figure 29. (a) Sketch of the attractor: the successive reflexions of an internal wave in a closed domain with one inclined wall converge toward a limited region, referred to as an attractor. (b) Laboratory experiment of the attractor in a tank stably-stratified with salted water. Horizontal lines of dye have been introduced at initial time, whose subsequent motions attest of the internal wave formation. To make the attractor more visible, the image of the flow at initial time has been substracted to that at a time where the attractor has formed. The localized presence of dye lines attest of localized fluid motions (from Maas et al., 1997).

Wunsch (1998) suggested that the breaking and mixing of the internal tide could fundamentally contribute to the global thermal equilibrium of the ocean: the cold bottom water masses in the thermohaline circulation indeed raise toward the surface through mixing; and mixing in the abyssal ocean, away from the boundaries, mostly occurs through internal wave breaking. Several questions therefore stem from this argument: what is the rate of conversion from the barotropic tide into the baroclinic tide? How and where does the internal tide lose energy? In particular, where does breaking occur? Since the conversion of energy is largest at continental slopes, we considered the generation of the internal tide at a continental slope.

The numerical modelling of this problem is not simple because the internal tide dynamics are nonlinear, non-hydrostatic, three-dimensional and a bathymetry is involved. Moreover, a continental slope typically extends from a depth of about 200m down to the abyssal plain, at 4000m. We are currently studying this problem with the code developed by John Marshall at MIT (Marshall et al., 1997) with the investigation of breaking and mixing as the main objective.

Figure 30a displays the internal tide velocity field for a linear hydrostatic computation performed by Gerkema (2002). This simpler context allows one to focus on the generation and propagation of the internal tide. The Brunt-Väisälä frequency is constant in this situation and equal to $2 \, 10^{-3}$ rad/s, the Coriolis frequency is that of mid-latitudes (10^{-4} rad/s) and the topography has a simple shape. Figure 30a has two remarkable features: the internal tide is generated at the shelf break and manifests itself as a single ray reflecting at the bottom and at the surface. Simple qualitative arguments can be provided to account for these features. Because the tide must jump over the slope, its velocity field acquires a vertical velocity over the slope which is largest on top of the slope. Let us consider a frame of reference attached to the barotropic tide. In this frame the topography has an oscillatory motion at the tidal frequency. This is analogous to the motion of an oscillating body in a stably-stratified medium already discussed. An internal gravity wave field is thus generated in the region of largest vertical velocity (if the oscillating frequency is comprised between the local values of N and f), whose energy propagates along a cross with angle set by the dispersion relation. In the present case, only parts of the cross can form because the body (the topography) is not spatially bounded, which propagate toward the abyssal plain and toward the shelf.

The inclusion of nonlinear non-hydrostatic effects strongly modify this regular pattern (Figure 30b). As discussed by Gerkema et al. (2005), an internal wave beam in an infinite medium is a solution of the fully nonlinear Boussinesq equations (see also Tabaei and Akylas, 2003). Hence, nonlinear effects are expected to occur where the beam meets a boundary. Figure 30b shows indeed that small-scale structures are observed in the generation region and where the beam meets the abyssal plain. One can explain the former observation by noticing that the beam is tangent to the slope in the generation region and is therefore aligned with the barotropic flow, so that forcing by the latter flow is optimum. In the reflexion region, higher harmonics are generated as a result of the superposition of the incident and reflected beams. As discussed in Section 6.1, the generation of a second harmonic, predicted theoretically by Dauxois and Young (1999), has been observed in a laboratory experiment of Peacock and Weidman (2005) as a result of the reflexion of an internal gravity beam onto a sloping boundary. No higher harmonics

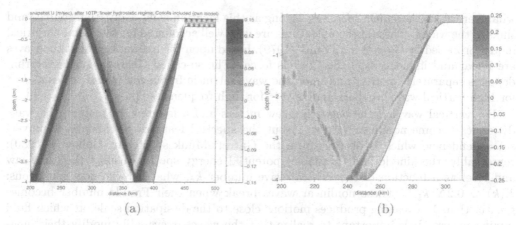

(a) (b)

Figure 30. (a) Horizontal velocity field at 10 tidal periods, using a linear hydrostatic numerical code (from Gerkema, 2002). (b) Same as (a), using the nonlinear non-hydrostatic MIT code.

are observed however in the experiment, as opposed to the present simulation, because the internal wave beam is not forced.

It follows that higher frequencies are generated once the beam reflects, and this is attested by frequency spectra of the kinetic and potential energy (Figure 31). The slope spectrum is close to -2 and is made of quasi-linear waves, suggesting that the spectra may be of the Garrett-Munk type (see next Section).

A practical application of this study is that two well-distinct regions appear to be turbulent, as further discussed in Gerkema et al. (2005), so that any parameterization of the internal tide may be seeked for independently in each region. Of course, this point should be complemented by simulations in a more realistic context.

7 Statistical properties of "wave turbulence"

As we recall it in Section 3.3, internal gravity waves transport the momentum and energy of their source, without any transport of matter as long as their dynamics are linear and non-dissipative (see, f.i., Andrews et al., 1987). The degree of nonlinearity depends upon the length scales involved and is quantified by the wave steepness (defined in Section 4.1).

In the ocean, $s \ll 1$ for vertical wavelengths comprised between a few tens and a few hundreds of meters and the wave dynamics are weakly nonlinear[5] (Garrett and Munk, 1979). Velocity and temperature spectra measured in the ocean at these scales remarkably display the same dependency on the vertical wavenumber and frequency

---

[5]At such scales, Coriolis effects are important and the Rossby number $Ro = A_u |\vec{k}|/f$ should actually be used in place of the Froude number. These large scales are characterized by $Ro \ll 1$. This implies that $Fr \ll 1$, since $Fr = Ro(f/N)$ and $f/N \ll 1$ in geophysical fluids. The Froude number becomes the relevant parameter at smaller scales.

whatever the measurement location, as long as this location is far away from sources and sinks of the waves. These spectral features are very well accounted for by a semi-empirical model proposed by Garrett and Munk (1979), based upon the assumption that the waves are linear and influenced by the Coriolis force. The so-called Garrett-Munk spectrum depends separately upon k_z and upon the wave intrinsic frequency Ω and scales as k_z^{-2} for high vertical wave numbers and as Ω^{-2} for high frequencies Ω.

For vertical wavelengths between a few meters and a few tens of meters, the wave dynamics become nonlinear ($Fr < 1$) but the spectral features still display a universal dependency, which is distinct from the Garrett-Munk spectrum (Holloway, 1980): remarkably, the kinetic and (available) potential energy spectra display the same law and level as a function of the vertical wave number k_z, whatever the flow conditions: $E(k) \simeq 0.2N^2 k_z^{-3}$. These nonlinear waves break when their Froude number becomes greater than 1. Breaking produces motions close to the dissipation scale, at which fluid mixing occurs. It is important to realize that the waves irreversibly modify their environment when they break, through a transport of mass both along the vertical direction (*i.e.* through fluid mixing[6]) and along the horizontal direction (as a mean flow is induced through deposition of momentum) (McIntyre and Norton, 1990).

The ubiquitous occurence of the latter spectrum is illustrated in Figure 32, through spectra of the available potential energy. In frame (a), data collected from oceanic measurements of the vertical temperature gradient are displayed as a function of k_z. Note that the spectrum is nearly flat for small k_z (the exponent is 0.1), corresponding to the Garrett-Munk spectrum. Frame (b) displays very high resolution measurements of the horizontal fluctuating velocity field in the stratosphere. The spectrum is of the

---

[6]Because of the existence of the restoring force of buoyancy along the vertical direction, no net vertical transport of mass can occur in a stably stratified fluid without mixing.

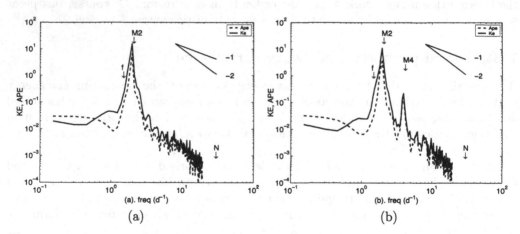

                  (a)   (b)

Figure 31. Frequency kinetic and potential energy spectra of the internal tide field (a) before reflexion of the wave beam; (b) after reflexion of the wave beam. The spectra are averaged over the vertical direction (from Gerkema et al., 2005).

Kolmogoroff type ($\sim k^{-5/3}$) for a vertical wavelength smaller than $\simeq$ 10m. In frame (c), the spectrum stems from a laboratory experiment of mixing of a density interface by an oscillating grid (Hannoun and List, 1988). For a certain parameter regime, internal gravity waves develop at the interface and break. In frame (d), the result from a two-dimensional (in a vertical plane) direct numerical simulation of a large scale breaking internal gravity wave in a constant N stratified fluid at rest is presented (Bouruet-Aubertot et al., 1996). The spectra are averaged over 14 Brunt-Väisälä periods (BVP) around the breaking time ($t \simeq 35$ BVP). An identical k_z^{-3} spectrum is obtained when the computation is performed in three dimensions (Carnevale et al., 2001).

8 Mixing by breaking internal gravity waves

The importance of mixing stems from the fact that small-scale processes, such as internal gravity waves, have an impact upon their environment only through irreversible processes, such as mixing. Hence, in large scale circulation models, these small-scale motions are usually represented through their mixing properties, via a turbulent diffusivity. Internal gravity waves deposit momentum when they break so that this effect should also be taken into account through a (isopycnal) transport coefficient.

It is not possible to ignore mixing from these small scale motions. As we already stressed it, in the ocean for instance, the cold bottom water masses raise through mixing and it is very likely that the thermohaline circulation, with typical time scale of one thousand years, would stop without mixing, with typical time scale of one minute. As well, in the atmosphere, the climate dynamics strongly depend upon mixing processes, which arise in convective clouds (Bony et al., 1995).

In this section, we only address basic concepts of mixing. A more extensive presentation along with measurements of mixing in a few laboratory and numerical experiments and in geophysical fluids can be found in Staquet (2004).

8.1 Basic concepts

Motions in a stably-stratified fluid are associated with kinetic and available potential energy (APE). The APE is partly reversibly converted into kinetic energy through the buoyancy flux, and partly irreversibly converted into background potential energy (BPE) through mixing (Lorenz, 1955; Thorpe, 1977; Winters et al., 1995). The BPE is the potential energy of the background density field $\rho_b(z)$, namely the density field that would exist in the fluid if it were brought to rest. The BPE is therefore the minimum potential energy of the fluid. The physical mechanism to account for the irreversible conversion from APE into BPE is simple: when fluid motions occur, fluid particles of different density are brought into contact and mix through molecular effects. This point can be discussed further. Mixing between two particles depends upon the time the particles are in contact. In other words, the Reynolds number of the fluid motion that advects the particles comes into play: if fluid particles are transported by large scale motions, with large Reynolds number, the diffusive time will be large relative to the advective time and mixing hardly occurs (but stirring does). By contrast, small scale motions will lead to more mixing. These ideas are illustrated in Staquet (2000) for the

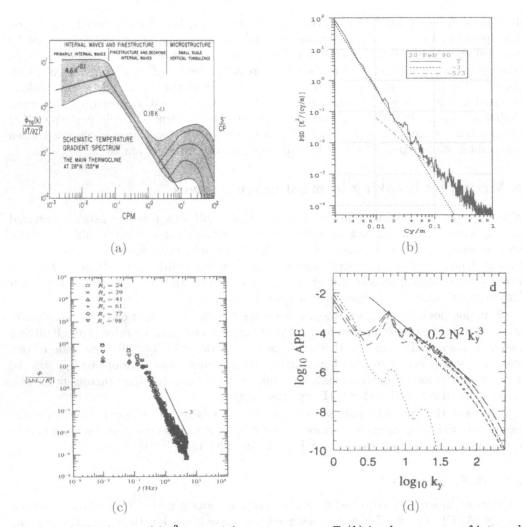

(a)

(b)

(c)

(d)

Figure 32. Evidence of k_z^{-3} potential energy spectra $E_p(k)$ in the presence of internal gravity wave breaking. a) Schematic spectra of vertical temperature gradients $k_z^2 E_p(k)$ in the oceanic thermocline (from Gregg, 1987). b) Stratospheric temperature spectra measured from instrumented balloons, showing a k_z^{-3} range with a Kolmogoroff $k_z^{-5/3}$ tail at high k_z (provided by F. Dalaudier). c) Normalized internal wave spectra, computed from fluoresceine concentration profiles in a laboratory experiment of mixing of a density interface by an oscillating grid. Δb is the buoyancy jump across the density interface, l_0 is an integral scale of turbulence and R_j is a Richardson number defined by $\Delta b\, l_0/u_0^2$, with u_0 being the rms of the horizontal velocity measured in a homogeneous fluid at the same distance from the oscillating grid as the density interface (from Hannoun and List, 1988). d) Two-dimensional direct numerical simulations of a large scale breaking internal gravity wave (k_y is the vertical wavenumber) (from Bouruet-Aubertot et al., 1996).

stably-stratified shear layer.

At the scale of the fluid system, mixing is quantified by the rate of change of the BPE. This rate of change is always positive, a consequence of the second law of thermodynamics. This rate is also the dissipation rate of the APE and will be referred to as the mixing rate ϵ_{APE}. The efficiency of mixing is characterized by the ratio of this rate of change to the amount of energy brought to the system per time unit. If a statistically steady state is reached, the amount of energy brought to the system per time unit is equal to the rate of energy dissipated per time unit in the system, namely $\epsilon_{APE} + \epsilon_{KE}$. The efficiency of mixing is therefore $\dfrac{\epsilon_{APE}}{\epsilon_{APE} + \epsilon_{KE}}$ and is referred to as the flux Richardson number in the oceanographic community (see Toole, 1998, for a discussion of mixing in the oceanographic context). The crucial point is therefore to compute ϵ_{APE}.

8.2 How to compute the mixing rate ϵ_{APE}?

An exact expression for the APE is presented in Holliday and McIntyre (1981), in terms of an expansion that holds for general $N(z)$. For constant N only the first term of the expansion remains, giving $APE = -\dfrac{g}{2d\overline{\rho}/dz} < \rho'^2 >$, where $< >$ denotes a volume average and $\overline{\rho}(z)$ is the horizontally averaged density profile. Note that for a constant N, the scale of the vertical gradient of the density profile is theoretically infinite so that this case corresponds to the linear limit. From a practical point of view, this expression of the APE is also used when the displacement of the fluid particles is small with respect to the scale of the vertical density gradient $d\overline{\rho}/dz$. In this case, from the linearized Boussinesq equations, the dissipation rate of APE is simply $-\kappa\dfrac{g}{d\overline{\rho}/dz} < |\nabla\rho'|^2 >$.

This method cannot be used when the particle displacements are not small with respect to the scale of the density gradient (but the Boussinesq approximation still holds). The stably-stratified shear layer is one example. In this case, Winters et al. (1995) proposed an exact analytical expression for ϵ_{APE}, based upon the background density profile. Hence, one has to compute this profile and two methods have been proposed for this purpose. The first method was proposed by Thorpe (1977) and extended to two and three dimensions by Winters et al. (1995). The background density profile is obtained by sorting adiabatically the density field so as to get a stable profile (note that the energy released by the sorting process is the APE). From a practical point of view, one sorts the values of the density field at each grid point so that a stable profile is reached; each plane of the numerical domain (supposed to be parallelepipedic), starting from the lowest altitude, is then filled with the heaviest particles and the density is averaged over each plane to get a monotonic density profile. Tseng and Ferziger (2001) proposed another method to access the background density profile, which is more efficient numerically than the sorting method and relies upon the probability density function of the density field.

As said above, these theoretical approaches have been developed to compute the mixing efficiency. Mixing is also quantified by a (diapycnal) turbulent diffusivity K_ρ, defined as $\epsilon_{APE}/g = -K_\rho\dfrac{d\rho_b}{dz}$. In the linear limit, $\overline{\rho} \simeq \rho_b$ for all times and one recovers the expression widely used in oceanography: $K_\rho = -\kappa \dfrac{< \overline{|\nabla \rho'|^2} >}{(d\bar\rho/dz)^2}$.

9 Conclusion

We tried in these Notes to broadly describe the very rich dynamics of internal gravity waves. The very study of those waves started in the mid-sixties, when Ball (1964) realized that two surface waves could resonantly interact with an internal gravity wave. Much has been learned about these waves, especially from a theoretical point of view, although the dispersion relation of inertia-gravity waves has just been verified experimentally (Peacock and Tabaei, 2005). And much remains to be done when natural media are considered. The sources of internal gravity waves in the atmosphere, other than the wind blowing over orography, are still the subject of active research. For instance, convective clouds at low latitudes generate such waves, which may interact in turn with the clouds. As well, the analysis of atmospheric data (FASTEX campaign) revealed that energetic large-scale low-frequency waves, with velocity comprised between 8 and 10 m/s, are emitted by the tropospheric jet (Plougonven et al., 2003). The important role of internal gravity waves in mixing the abyssal water masses has been conjectured but not proved. And the role of these waves in the dynamics of the radiative zone of the Sun is totally unknown. From a modelling point of view, weakly nonlinear theories have been developed to account for the statistical properties of those waves (see Staquet and Sommeria, 2002). But, from a practical point of view, no reliable parameterization of the waves exists, which would relate the transport coefficients they induce to the dynamical parameters of the large scale motions which create them. Such parameterizations are crucially needed in general circulation models of the atmosphere and the ocean.

10 Acknowledgments

The author thanks V. Armenio and S. Sarkar for their invitation to the CISM Summer School on "Environmental stratified flows" in Udine, for very fruitful discussions and for providing the opportunity to write these lecture notes.

Bibliography

D. G. Andrews, J. R. Holton, and C. B. Leovy. *Middle atmosphere dynamics*. Academic Press, 1987.

S. I. Badulin and V. I. Shrira. On the irreversibility of internal-wave dynamics due to wave trapping by mean flow inhomogeneities. part 1. local analysis. *J. Fluid Mech.*, 251:21–53, 1993.

S. I. Badulin, V. I. Shrira, and L. S. Tsimring. The trapping and vertical focusing of internal waves in a pycnocline due to the horizontal inhomogeneities of density and currents. *J. Fluid Mech.*, 158:199–218, 1985.

S. I. Badulin, V. M. Vasilenko, and N. N. Golenko. Transformation of internal waves in the equatorial lomonosov current. *Atmospheric and Oceanic Physics.*, 26(2):110–117, 1990.

K. Ball. Energy transfer between external and internal gravity waves. *J. Fluid Mech.*, 19:465–478, 1964.

J. Barat and C. Cot. Wind shear rotary spectra in the atmosphere. *Geophys. Res. Letters.*, 19(2):103–106, 1992.

A. Y. Basovich and L. S. Tsimring. Internal waves in a horizontally inhomogeneous flow. *J. Fluid Mech.*, 142:233–49, 1984.

D. Benielli and J. Sommeria. Excitation and breaking of internal gravity waves by parametric instability. *J. Fluid Mech.*, 374:117–144, 1998.

S. Bony, J.-P. Duvel, and H. Le Treut. Observed dependence of the water vapor and clearsky greenhouse effect on sea-surface temperature: comparison with climate warming experiments. *Climate Dynamics*, 11:307–320, 1995.

J. R. Booker and F. P. Bretherton. The critical layer for internal gravity waves in a shear flow. *J. Fluid Mech.*, 27:513–539, 1967.

P. Bouruet-Aubertot, J. Sommeria, and C. Staquet. Stratified turbulence produced by internal wave breaking: two-dimensional numerical experiments. *Dyn. Atmos. Oceans*, 23:357, 1996.

F. P. Bretherton. The propagation of groups of internal gravity waves in a shear flow. *Quart. J. Roy. Met. Soc.*, 92:466–480, 1966.

F. P. Bretherton. Propagation in slowly varying waveguides. *Proc. Roy. Soc. A*, 302: 555–576, 1968.

G. F. Carnevale, M. Briscolini, and P. Orlandi. Buoyancy- to inertial-range transition in forced stratified turbulence. *J. Fluid Mech.*, 427:205 – 239, 2001.

S. Chandrashekar. *Hydrodynamic and hydromagnetic stability.* Oxford: Clarendon press, 1961.

B. Cushman-Roisin. *Introduction to Geophysical Fluid Dynamics.* Prentice-Hall, 1994.

F. Dalaudier, C. Sidi, M. Crochet, and J. Vernin. Direct evidence of "sheets" in the atmospheric temperaturefield. *J. Atmosph. Sciences*, 51:237–248, 1994.

T. Dauxois and W. R. Young. Near-critical reflection of internal waves. *J. Fluid Mech.*, 390:271–295, 1999.

P. G. Drazin and W. H. Reid. *Hydrodynamic Stability.* Cambridge University Press, Cambridge, 1981.

P.G. Drazin. On the parametric instability of an internal gravity wave. *Proc. R. Soc. Lond. A*, 356:411, 1977.

T. J. Dunkerton. Shear instability of internal inertia-gravity waves. *J. Atmos. Sciences*, 54:1628, 1997.

N. R. E. Edwards and C. Staquet. Focusing of an inertia-gravity wave packet by a baroclinic shear flow. *Dynamics of Atmospheres and Oceans*, In press, 2005.

A. Eliassen and E. Palm. On the transfer of energy in stationary mountain waves. *Geofys. Publ.*, 22(3):1–23, 1961.

D. M. Farmer and L. Armi. Stratified flow over topography: the role of small scale entrainment and mixing in flow establishment. *Proc. Roy. Soc., London A*, 455:3221–3258, 1999.

D.C. Fritts, J.R. Isler, and O. Andreassen. Gravity wave breaking in two and three dimensions. 2. Three-dimensional evolution and instability structure. *J. Geophys. Res.*, 99(D4):8109–8123, 1994.

C. Garrett and W. Munk. Internal waves in the ocean. *Ann. Rev. Fluid Mech.*, 11: 339–369, 1979.

T. Gerkema. Application of an internal-tide generation model to baroclinic spring-neap cycles. *J. Geophys. Res.*, 107(C9):3124, 2002.

T. Gerkema, C. Staquet, and P. Bouruet-Aubertot. Nonlinear effects in internal-tide beams, and mixing. *Ocean modelling*, Accepted for publication, 2005.

M. Gregg. Diapycnal mixing in the thermocline: a review. *J. Geophys. Res.*, 92(C5): 5249–5286, 1987.

I. A. Hannoun and E. J. List. Turbulent mixing at a shear-free density interface. *J. Fluid Mech.*, 189:211–234, 1988.

K. Hasselmann. On the non-linear energy transfer in a gravity-wave spectrum. Part 1: General theory. *J. Fluid Mech.*, 12:481–500, 1962.

K. Hasselmann. A criterion for nonlinear wave stability. *J. Fluid Mech.*, 30:737, 1967.

P. Hazel. Numerical studies of the stability of inviscid stratified shear flows. *J. Fluid Mech.*, 51:39–61, 1972.

G. J. F. van Heijst, R. C. Kloosterziel, and C. W. M. Williams. Laboratory experiments on the tripolar vortex in a rotating fluid. *J. Fluid Mech.*, 225:301–331, 1991.

D Holliday and M. E. McIntyre. On potential energy density in an incompressible stratified fluid. *J. Fluid Mech.*, 107:221–225, 1981.

G. Holloway. Oceanic internal waves are not weak waves. *J. Phys. Oceanogr.*, 10:906–914, 1980.

G. Holloway and D. Ramsden. Theories of internal wave interaction and stably stratified turbulence: testing against numerical simulations. In J. C. Nihoul and B.M. Jamard, editors, *Small-scale Turbulence and Mixing in the Ocean*, 'Aha Huliko'a Hawaiian Winter Workshop, pages 363–377. Elsevier, New-York, 1988.

L. N. Howard. Note on a paper of john w. miles. *J. Fluid Mech.*, 10:509–512, 1961.

Y. A. Ivanov and Y. G. Morozov. Deformation of internal gravity waves by a stream with horizontal shear. *Okeanologie*, 14:457–461, 1974.

G. N. Ivey, K. B. Winters, and I. P. D. De Silva. Turbulent mixing in a sloping benthic boundary layer energized by internal waves. *J. Fluid Mech.*, 418:59–76, 2000.

W. J. Jones. Propagation of internal gravity waves in fluids with shear flow and rotation. *J. Fluid Mech.*, 30:439–449, 1967.

W. J. Jones. Reflexion and stability of waves in stably stratified fluids with shear flow: a numerical study. *J. Fluid Mech.*, 34:609–624, 1968.

P. Klein, B. L. Hua, and X. Carton. Emergence of cyclonic structures due to the interaction between near-inertial oscillations and mesoscale eddies. *Quart. J. Roy. Met. Soc.*, 129:1–20, 2003.

J. Klostermeyer. Parametric instabilities of internal gravity waves in boussinesq fluids wit large reynolds numbers. *Geophys. Astrophys. Fluid. Dyn.*, 26:85, 1983.

J. Klostermeyer. Two and three-dimensional parametric instabilities in finite amplitude internal gravity waves. *Geophys. Astrophys. Fluid Dyn.*, 61:1, 1991.

C. G. Koop. A preliminary investigation of the interaction of internal gravity waves with a steady shearing motion. *J. Fluid Mech.*, 113:347–386, 1981.

C. G. Koop and B. McGee. Measurements of internal gravity waves in a continuously stratified shear flow. *J. Fluid Mech.*, 172:453–480, 1986.

C. R. Koudella. *Ondes internes de gravité : instabilités, déferlement et vorticité potentielle.* PhD thesis, École Normale Supérieure de Lyon, France, 1999.

C. R. K. Koudella and C. Staquet. Instability mechanisms of a two-dimensional progressive internal gravity wave. *J. Fluid Mech.*, submitted, 2005.

P. H. Leblond and L. A. Mysak. *Waves in the Ocean.* Number 20 in Elsevier Oceanography Series. Elsevier Scientific Publishing Company, 1978.

M. P. Lelong and T. J. Dunkerton. Inertia-gravity wavesbreaking in three dimensions: 2. convectively unstable waves. *J. Atmos. Sci.*, 55(15):2489, 1998.

M. Lesieur. *Turbulence in Fluids.* Kluwer Academic Publishers, 3 edition, 1997.

M.J. Lighthill. *Waves in Fluids.* Cambridge University Press, 1978.

P.N. Lombard and J.J. Riley. Instability and breakdown of internal gravity waves. I. linear stability analysis. *Phys. Fluids*, 8(12):3271, 1996.

E.N. Lorenz. Available potential energy and the maintenance of the general circulation. *Tellus*, 7:157–167, 1955.

F. Lott. Comparison between the orographic response of the ecmwf model and the pyrex 1990 data. *Quarterly Journal of the Royal Meteorological Society*, 121:1323–1348, 1995.

L. R. M. Maas, D. Benielli, J. Sommeria, and F.-P. A. Lam. Observation of an internal wave attractor in a confined, stably stratified fluid. *Nature*, 388:557, 1997.

L. R. M. Maas and F.-P. A. Lam. Geometric focusing of internal waves. *J. Fluid Mech.*, 300:1–41, 1995.

J. Marshall, A. Adcroft, C. Hill, L. Perelman, and C. Heisey. A finite-volume, incompressible navier-stokes model for studies of the ocean on parallel computers. *J. Geophys Res.*, 102(C3):5753–5766, 1997.

C. H. McComas and F. P. Bretherton. Resonant interactions of oceanic internal waves. *J. Geophys. Res.*, 82:1397–1411, 1977.

A.D. McEwan. Degeneration of resonantly-excited standing internal gravity waves. *J. Fluid Mech.*, 50:431–448, 1971.

A.D. McEwan and R.A. Plumb. Off-resonant amplification of finite internal wave packets. *Dyn. Atmos. Ocean*, 2:83–105, 1977.

A.D. McEwan and R. M. Robinson. Parametric instability of internal gravity waves. *J. Fluid Mech.*, 67(4):667–687, 1975.

M. E. McIntyre. The stratospheric polar vortex and sub-vortex : fluid dynamics and mid-latitude ozone loss. *Phil. Trans. Roy. Soc. London*, 352:227–240, 1995.

M. E. McIntyre. On global-scale atmospheric circulations. In G. K. Batchelor, H. K. Mofatt, and M. G. Worster, editors, *Perspectives in Fluid Dynamics*, pages 557–624. Cambridge University Press, 2000.

M. E. McIntyre and W. A. Norton. Dissipative wave-mean interactions and the transport of vorticity or potential vorticity. *J. Fluid Mech.*, 212:403–435, 1990.

E.E. McPhee-Shaw and E. Kunze. Boundary-layer intrusions from a sloping bottom: A mechanism for generating intermediate nepheloid layers. *J. Geophys. Res.*, 107: DOI:10.1029/2001JC000801, 2002.

R.P. Mied. The occurence of parametric instabilities in finite-amplitude internal gravity waves. *J. Fluid Mech.*, 78:763–784, 1976.

T. Miyazaki and K. Adachi. Short-wavelength instabilities of waves in rotating stratified fluids. *Phys. Fluids*, 10(12):3168–3177, 1998.

D. E. Mowbray and B. S. H. Rarity. A theoretical and experimental investigation of the phase configuration of internal waves of small amplitude in a density stratified liquid. *J. Fluid Mech.*, 28:1–16, 1967.

W. Munk and C. Wunsch. Abyssal recipes ii: energetics of tidal and wind mixing. *Deep-Sea Research I*, 45:1977, 1998.

W. H. Munk. A survey of internal waves and small scale processes. In B. A. Warren and C. Wunsch, editors, *Evolution of Physical Oceanography*, pages 264–291. MIT Press, Cambridge, Mass., 1981.

P. Muller, G. Holloway, F. Henyey, and N. Pomphrey. Nonlinear interactions among internal gravity waves. *Rev. Geophys.*, 24(3):493–536, 1986.

D. J. Olbers. The propagation of internal waves in a geostrophic current. *J. Phys. Oceanogr.*, 11:1224–1233, 1981.

M. C. Ollers, L. P. J. Kamp, F. Lott, P. F. J. van Velthoven, H. M. Kelder, and F. W. Sluijter. Propagation properties of inertia-gravity waves through a barotropic shear layer and application to the antarctic polar vortex. *Quartely Journal of the Royal Meteorological Society*, 129:2495–2911, 2003.

T. Peacock and A. Tabaei. Visualization of nonlinear effects in reflecting internal wave beams. *Phys. Fluids*, Accepted for publication, 2005.

T. Peacock and P. Weidman. The effect of rotation on conical wave beams in a stratified fluid. *Expts. Fluids*, To appear, 2005.

O. M. Phillips. On the dynamics of unsteady gravity waves of finite amplitude, i. *J. Fluid Mech.*, 9:193–217, 1960.

O. M. Phillips. Wave interactions - the evolution of an idea. *J. Fluid Mech.*, 106:215–227, 1981.

O.M. Phillips. *The dynamics of the upper ocean*. Cambridge University Press, 1966.

R. Plougonven, H. Teitelbaum, and V. Zeitlin. Inertia-gravity wave generation by the tropospheric mid-latitude jet as given by the fastex radiosoundings. *Journal of Geophysical Research*, 108(D21):4686, 2003.

J. M. Rees, J. C. W. Denholm-Price, J. C. King, and P. S. Anderson. A climatological study of internal gravity waves in the atmospheric boundary layer overlying the brunt ice shelf, antarctica. *Journal of the Atmospheric Sciences*, 57:511–526, 2000.

E. Schatzman. Diffusion process produced by random internal waves. *J. Fluid Mech.*, 322:355–382, 1996.

D. G. Schowalter, J. C. Lasheras, and C. W. van Atta. A study of streamwise vortex structures in a stratified shear layer. *J. Fluid Mech.*, 281:247–291, 1994.

D. N. Slinn and J. J. Riley. Turbulent dynamics of a critically reflecting internal gravity wave. *Theor. Comp. Fluid Dyn.*, 11:281–303, 1998.

L. J. Sonmor and G. P. Klaassen. Higher-order resonant instabilities of internal gravity waves. *J.Fluid Mechanics*, 324:1–23, 1996.

C. Staquet. Mixing in a stably stratified shear layer: two- and three-dimensional numerical experiments. *Fluid Dyn. Research*, 27:367–404, 2000.

C. Staquet. Gravity and inertia-gravity internal waves : breaking processes and induced mixing. *Surveys in Geophysics*, 25(3-4):281–314, 2004.

C. Staquet and G. Huerre. On transport across a barotropic shear flow by breaking inertia-gravity waves. *Physics of Fluids*, 14(6):1993–2006, 2002.

C. Staquet and J. Sommeria. Mixing by breaking internal gravity waves: from instabilities to turbulence. *Annual Reviews Fluid Mech*, 34:559–593, 2002.

R. B. Stull. *Boundary layer meteorology*. Kluwer Academic Publisher, 1988.

B. R. Sutherland. Finite-amplitude internal wave-packet dispersion and breaking. *J. Fluid Mech.*, 429:343–380, 2001.

A. Tabaei and T. R. Akylas. Nonlinear internal gravity wave beams. *J. Fluid Mech.*, 482:141–161, 2003.

A. Tabaei, T. R. Akylas, and K. G. Lamb. Nonlinear effects in reflecting and colliding internal wave beams. *J. Fluid Mech.*, 526:217–243, 2005.

S. Talon, P. Kumar, and J.-P. Zahn. Angular momentum extraction by gravity waves in the sun. *Astrophys. J. Letters*, 574:L175–L178, 2002.

S. A. Thorpe. Turbulence and mixing in a scottish loch. *Phil. Trans. R. Soc. Lond. A*, 286:125–181, 1977.

J. Toole. Turbulent mixing in the ocean: intensity, causes and consequences. In E. Chassignet and J. Verron, editors, *Ocean modelling and parameterization*, Proccedings of Les Houches Winter School, pages 171–190. Kluwer Academic Publishers, 1998.

Y. H. Tseng and J. H. Ferziger. Mixing and available potential energy in stratified flows. *Physics Fluids*, 33(2):1281–1293, 2001.

A. Veysseyre, K. Moutard, C. Ferrari, K. Van de Velde, C. Barbante, G. Cozzi, G. Capodaglio, and C. Boutron. Heavy metals in fresh snow collected at different altitudes in the chamonix and maurienne valleys, french alps: initial results. *Atmospheric Environment*, 35(2):415–425, 2000.

K. B. Winters and E. A. D'Asaro. Two-dimensional instability of finite amplitude internal gravity wave packets near a critical level. *J. Geophys. Res.*, 94(C9):12,709–19, 1989.

K. B. Winters and E. A. D'Asaro. Three-dimensional wave breaking near a critical level. *J. Fluid Mech.*, 272:255–284, 1994.

K. B. Winters, P. N. Lombard, J. J. Riley, and E. A. D'Asaro. Available potential energy and mixing in density-stratified fluids. *J. Fluid Mech.*, 289:115–128, 1995.

K. B. Winters and J. J. Riley. Instability of internal waves near a critical level. *Dyn. Atmos. Oceans*, 16:249–278, 1992.

M. G. Wurtele, A. Datta, and R. D. Sharman. The propagation of gravity-inertia waves and lee waves under a critical level. *J. Atmos. Sciences*, 53(11):1505, 1996.

O. Zikanov and D. N. Slinn. Along-slope current generation by obliquely incident internal waves. *J. Fluid Mech.*, 445:235 – 261, 2001.

Chapter 3: Prototypical examples of stratified shear flow

Sutanu Sarkar

University of California at San Diego

Abstract. Stratification effects on turbulence are examined in some fundamental shear flows. The differences between unbounded flows and those with walls are indicated. The role of the gradient Richardson number is assessed. Detailed results on turbulence energetics, transport and mixing are presented.

1 Introduction

Mean shear produces instabilities and turbulence while stratification generally dampens the intensity of turbulent velocity fluctuations while allowing internal waves. The competition between mean shear and stratification in determining the fluctuating motion and the associated buoyancy flux, momentum flux and molecular mixing has been studied in a variety of flows. The present review is limited to prototypical shear flows that we have investigated using direct numerical simulation (DNS) or large eddy simulation (LES). Other perspectives on the subject are available in the reviews of Lin and Pao (1979); Hopfinger (1987); Fernando (1991); Riley and Lelong (2000); Peltier and Caulfield (2003).

Vertical shear is the type of shear that is most common in practice and has been widely studied in the past. Section 2 shows results from the simplest possible case, that with constant mean shear and constant stratification. Here, there is only a single parameter, the gradient Richardson number, that measures the strength of stratification relative to shear, and furthermore its value is constant. Section 3 is a discussion of stratified channel flow that serves to illustrate stratification effects on wall-bounded turbulence. The gradient Richardson number is no longer constant and, besides, other Richardson number definitions are possible. The importance of the thermal boundary condition, adiabatic in the oceanic boundary layer in contrast to constant heat flux or constant temperature in the atmospheric boundary layer, is also discussed. There are instances when the mean shear is horizontal: in oceanic and atmospheric flows around topography, at the edges of energetic ocean currents, coastal fronts, river outflows, and behind self-propelled bodies. Section 4 presents results from the the simplest example of stratified horizontal shear flow, that when the mean shear is constant. The shear layer with inflectional mean shear is discussed in section 5 to illustrate the role of buoyancy in the dynamics of horizontal eddying motions and vortices that can be seen in oceans.

2 Uniform Vertical shear and Uniform Stratification

The simplest shear flow where the competing effects of shear and stratification may be studied is the situation with a uniform gradient of the ambient density S_ρ (therefore, constant Brunt-Väisälä frequency N) and uniform mean shear rate S. During the past two decades, both laboratory and numerical simulations of this flow have been performed. In the numerical investigations, the temporal evolution of an initial field of velocity fluctuations is simulated with imposed mean density and velocity profiles as shown by Figure 1. The turbulence is statistically homogeneous but its statistics evolve as a function of time, t. In the laboratory studies, the linear profiles of Figure 1 are imposed at the inflow on grid turbulence and the flow evolves spatially in the streamwise, x- direction. The turbulence is statistically steady and spatially homogeneous in all directions except the streamwise one. The connection between streamwise evolution measured in the laboratory and temporal evolution measured in the numerical simulations is made by employing $x = Ut$ where U is the centerline flow velocity in the laboratory.

The primary non-dimensional parameters that determines the flow evolution are the Reynolds number, the gradient Richardson number, the shear number, and the Prandtl number. The Reynolds number is usually given in the form $Re_\lambda = q\lambda/\nu$ that uses the Taylor microscale, λ, and the velocity fluctuation, $q = \sqrt{\overline{u_i u_i}}$. The Taylor microscale is defined by $\epsilon = 5\nu q^2/\lambda^2$. The gradient Richardson number measures the strength of the stratification relative to shear and is given by $Ri_g = N^2/S^2$, the square of the ratio of the Brunt-Väisälä frequency $N = \sqrt{(-g/\rho_0)S_\rho}$ and the mean shear rate S. The shear number, SK/ϵ, measures the ratio of shear to a frequency K/ϵ characterizing the energy-containing scales of turbulence. The Prandtl number, ν/α, is the ratio of molecular to thermal diffusivities. During the evolution of the flow, Ri_g and Pr are constant, the value of Re_λ increases with time and the value of shear number, SK/ϵ, also evolves from its initial value.

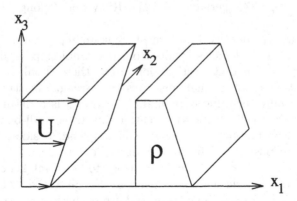

Figure 1. Sketch of the mean velocity and density in flow with uniform shear and stratification. Reproduced with permission from Jacobitz et al. (1997)

In uniform shear flow without stratification, it is found that K and ϵ grow exponentially with time (streamwise coordinate in the laboratory) with a well-defined growth rate, Rogers et al. (1989); Rohr et al. (1988); Tavoularis and Karnik (1989). The structural characteristics such as Reynolds stress anisotropies and the shear number approach constant values. An important result of the stratified flow studies is that the growth of K progressively decreases with increasing Ri_g as shown in Figure 2. A critical value of the Richardson number Ri_{cr} can be found for which the turbulent kinetic energy K stays constant in time, grows in time for $Ri < Ri_{cr}$, and decays in time for $Ri > Ri_{cr}$. The knowledge of the critical Richardson number is therefore important as it divides the asymptotic fate of the non-linearly evolving turbulence into growth or decay. There is an analogous situation in the evolution of linearized disturbances where it is found that a flow with arbitrary profiles, $U(z)$ and $\rho(z)$, is stable if $Ri > 1/4$ everywhere in the flow, Miles (1961); Howard (1961). The first direct numerical simulations of homogeneous turbulence in a stratified shear flow were performed by Gerz et al. (1989). In their investigation shear periodic boundary conditions were used in combination with a finite difference/spectral collocation method. The simulations were performed on a grid with 64^3 points. The Richardson number was varied over the range $0 \leq Ri \leq 1$. Gerz et al. (1989) found that the evolution of the turbulent kinetic energy K depends strongly on the Richardson number Ri. The initial energy partition η, the ratio of the potential to kinetic energy, was found to lead to different initial transients but not to influence the turbulence evolution over a long period of time. These authors also investigated the occurrence of counter gradient heat fluxes which were more prominent for large Richardson numbers and high molecular Prandtl numbers. Holt et al. (1992) investigated the turbulence evolution in a homogeneous stably stratified shear flow using purely periodic boundary conditions in combination with a spectral collocation method. The simulations were performed on a grid with 128^3 points. The simulations covered a parameter range $0 \leq Ri \leq 1$ and $20 \leq Re_\lambda \leq 100$. They divided the turbulence evolution into shear or buoyancy dominated regimes for the Richardson number smaller or larger than the transitional value Ri_t for which the vertical density flux vanishes. Simulations with constant turbulent kinetic energy K were found to lie in the shear dominated regime, that is $Ri_{cr} < Ri_t$. These authors also investigated the influence of the initial energy partition η and of the molecular Prandtl number Pr and agreed with the conclusions of Gerz et al. (1989). In addition the dependence of the turbulence evolution on the Reynolds number Re_λ was addressed. It was found that the critical Richardson number Ri_{cr} increases with the Reynolds number Re_λ. H.-J. Kaltenbach and Schumann (1994) performed large eddy simulations to investigate the turbulent transport in a homogeneous stably stratified shear flow. They observed that vertical overturning and mixing is suppressed when the Froude number $F = q/Nl$ (1 is an integral length scale of the velocity, and q is the root-mean-square velocity) is less than a critical value of about 0.3. Jacobitz et al. (1997) reconsidered the problem using DNS in order to systematically study the effect of the initial shear number, SK/ϵ, that had not been studied before, as well as Ri_g and Re_λ. It was shown that the growth rate of turbulent kinetic energy initially increased with increasing initial SK/ϵ and then decreased with the later decrease attributed to the persistence of linear (rapid distortion theory) effects. The critical Richardson number was found to vary in the range, $0.04 < Ri_{cr} < 0.17$ owing to its strong dependence on

Reynolds and shear numbers. Jacobitz and Sarkar (1999a) extended the initial Reynolds number to $Re_\lambda = 89$ and initial shear number to $SK/\epsilon = 50$ by using a $512 \times 256 \times 256$ grid. These authors found that the value of SK/ϵ required for the persistence of linear effects that inhibits the TKE growth increases with Reynolds number. Jacobitz and Sarkar (1999a) also examined the effect of stratification on energy spectra and spectral transport. Shih et al. (2000) reconsidered the shear number effect, also finding an effect at moderate values of $Re_\lambda = 44$ similar to Jacobitz and Sarkar (1999a); however, their series at $Re_\lambda = 89$ did not extend to a sufficiently large value of SK/ϵ for the shear number effect to be important.

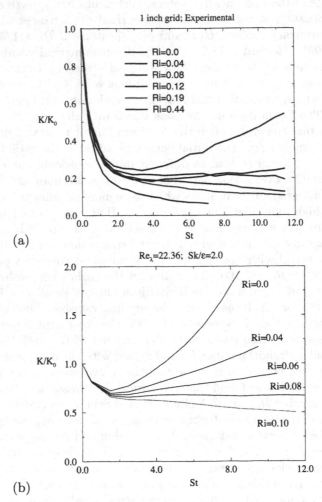

Figure 2. Evolution of TKE: (a) Laboratory result, (b) DNS result.

Rohr et al. (1988) performed the first successful experiments of homogeneous turbulence in a stably stratified shear flow in a salt-stratified water channel. A variation of the Richardson number Ri led to constant vertical velocity fluctuations u_3^{rms} for $Ri = Ri_{cr} = 0.25 \pm 0.05$, growing u_3^{rms} for $Ri < Ri_{cr}$, and decaying u_3^{rms} for $Ri > Ri_{cr}$. Piccirillo and VanAtta (1997) investigated the turbulence evolution in a homogeneous stably stratified shear flow using a thermally stratified wind tunnel, in which they varied the Reynolds number Re_λ over a small range by using a variety of different turbulence generating grids. They found a decrease of the critical Richardson number Ri_{cr} with increasing grid size and thus increasing Reynolds number Re_λ. Holt et al. (1992), on the other hand, found an increase of Ri_{cr} with increasing Reynolds number Re_λ. Jacobitz et al. (1997) resolved this controversy by showing that the apparent disagreement between the experiment of Rohr et al. (1988) and DNS of Holt et al. (1992) was a consequence of the initial SK/ϵ which took different values in the two studies. To summarize, for moderate values of the shear number and sufficiently large Reynolds number, the observed values of Ri_{cr} in the range $0.18 < Ri_{cr} < 0.25$ over all the DNS and laboratory studies performed to date are consistent with the linear analysis result that $Ri_g > 0.25$ is a sufficient condition for stability. In the remainder of this section on uniform shear flow, we give salient features of the work by Jacobitz et al. (1997); Jacobitz and Sarkar (1999a,b).

2.1   Numerical algorithm

The numerical scheme uses a spectral collocation method for the spatial discretization. To compute spatial partial derivatives, the dependent variables are transformed into Fourier space using the Fast Fourier Transformation algorithm, multiplied with the corresponding wavenumbers, and transformed back into physical space. Second derivatives are computed by applying this method successively. The non-linear terms are computed in physical space. The solution is advanced in time using a third order Runge-Kutta method. During the time advancement the coordinate system in the moving frame of reference becomes more and more skewed. Following a method originally devised by Rogallo (1981), the coordinate system is reoriented from $+45°$ to $-45°$ using the periodic structure of the dependent variables. The regridding procedure produces aliasing errors. These errors are controlled by truncation of the affected modes ($2k_{max}/3 < k < k_{max}$) before and after the regridding procedure. Jacobitz et al. (1997) performed simulations with 128^3-244^3 points while Jacobitz and Sarkar (1999a) used up to $512 \times 256 \times 256$ points.

The spectra of the fluctuations have to be defined carefully to minimize initial transients and allow a parameterization of the flow evolution based on the initial values of the parameters. During the early years of DNS, a random number generator was used to produce the initial fluctuating fields which followed a specified initial energy spectrum and fulfilled the continuity equation, but contained no phase information and had no initial spectral transfer. It is now accepted that isotropic turbulence fields, that satisfy the Navier-Stokes equations, obtained from a separate initialization simulation provide much better initial fields for the shear flow DNS. Besides, such initial conditions also compare better with those found in laboratory experiments, where nearly isotropic conditions develop behind the turbulence generating grid before the effects of shear and stratification become important. The initial value of the Reynolds number of the initialization simulation is chosen such that the final value matches the target initial Reynolds number of the actual simulation. The initialization simulation is advanced for about one eddy-turnover in time and thus beyond the initial transient. During this time the skewness of $\partial u/\partial x$, which is a measure of the spectral transport, increases from zero to a final value $Sk = -0.46$. This value is in good agreement with other experimental and numerical results of $-0.5 < Sk < -0.4$, Lesieur (1997).

2.2   TKE evolution

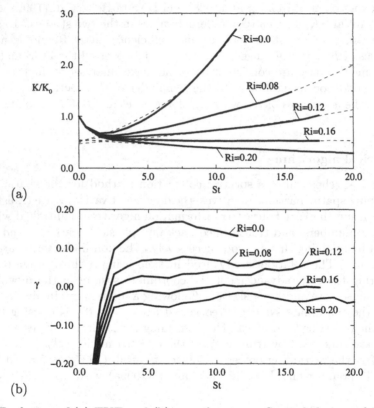

(a)

(b)

Figure 3. Evolution of (a) TKE and (b) growth rate, γ. Several choices of Ri are shown while the initial values of the Reynolds number $Re_\lambda = 44.72$ and the shear number $SK/\epsilon = 2.0$ are fixed. The dashed line shows the exponential approximation of Eq. 2.10. Reproduced with permission from Jacobitz et al. (1997).

To derive the non-dimensional parameters governing the turbulence evolution, the transport equation for the turbulent kinetic energy K is scaled. Let q be the characteristic velocity scale, l the characteristic length scale and r the characteristic density scale. A characteristic time scale for the turbulence evolution is then given by $\tau = l/q$. In addition the Taylor microscale λ is introduced as a derivative length scale by $\epsilon = 5\nu q^2/\lambda^2$. The terms of the transport equation for the kinetic energy K are divided by the dissipation ϵ:

$$\underbrace{\frac{1}{\epsilon}\frac{d}{dt}K}_{1} = \underbrace{\frac{P}{\epsilon}}_{2} + \underbrace{\frac{B}{\epsilon}}_{3} - 1 \,. \tag{2.1}$$

Recall that the turbulent production is

$$P = -S\overline{u_1 u_3} \,,$$

the buoyancy flux is

$$B = -\frac{g}{\rho_0}\overline{u_3 \rho} \,,$$

and the turbulent dissipation rate is

$$\epsilon = \nu \overline{\frac{\partial u_i}{\partial x_k}\frac{\partial u_i}{\partial x_k}} \,.$$

Note, that the buoyancy flux is sometimes (including our earlier papers) defined without the minus sign. Eq. (2.1) is scaled as follows:

$$O(\text{term 1}) = Re_\lambda \frac{\lambda}{l} \tag{2.2}$$

$$O(\text{term 2}) = \frac{SK}{\epsilon} \tag{2.3}$$

$$O(\text{term 3}) = Ri\frac{L_e}{l}(\frac{SK}{\epsilon})^2\frac{1}{Re_\lambda}\frac{l}{\lambda} \tag{2.4}$$

Here $Re_\lambda = q\lambda/\nu$ denotes the Reynolds number based on the Taylor microscale, $Ri = (gS_\rho)/(\rho_0 S^2)$ the Richardson number, and $L_e = r/S_\rho$ the Ellison scale. The kinetic energy is $K = O(q^2)$. Therefore, the non-dimensional parameters governing the turbulence evolution for this low Reynolds number scaling are the Richardson number Ri, the Taylor microscale Reynolds number Re_λ and the shear number SK/ϵ. In addition, the initial conditions influence the turbulence evolution by the length scale ratios λ/l of the velocity fields and L_e/l of the density field.

For high Reynolds numbers, the dissipation varies as $\epsilon = O(q^3/l)$ or equivalently $O(l/\lambda) = Re_\lambda$. This simplifies the scaling to:

$$O(\text{term 1}) = 1 \tag{2.5}$$

$$O(\text{term 2}) = \frac{SK}{\epsilon} \tag{2.6}$$

$$O(\text{term 3}) = Ri\frac{L_e}{l}(\frac{SK}{\epsilon})^2 \tag{2.7}$$

Therefore, the non-dimensional parameters governing the turbulence evolution for the high Reynolds number scaling are the Richardson number Ri, the shear number SK/ϵ, and the ratio L_e/l only.

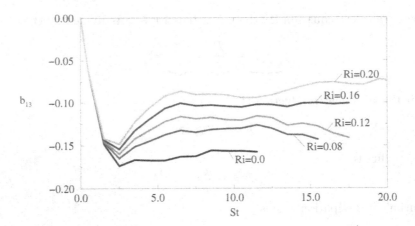

Figure 4. Evolution of b_{13} as a function of the Richardson number Ri; initial $Re_\lambda = 44.72$ and initial $SK/\epsilon = 2.0$. Reproduced with permission from Jacobitz et al. (1997).

All simulations show qualitatively the same dependence on the Richardson number Ri; one set of simulations with the initial values $Re_\lambda = 44.72$ and $SK/\epsilon = 2.0$ is discussed here. Figure 3a shows the evolution of the turbulent kinetic energy K as a function of the non-dimensional time St with the Richardson number Ri as the variable parameter. After an initial decay K grows for small Richardson numbers and decays for high Richardson numbers. This makes it possible to define a critical Richardson number Ri_{cr} when K is roughly constant.

The initial decay in all cases is due to the isotropic initial conditions that cause the production term $P = -S\overline{u_1 u_3}$ to be initially zero. During the initial phase the production term grows. The initial decay is also observed in the DNS of Gerz et al. (1989) and Holt et al. (1992), as well as the experimental studies of Rohr et al. (1988) and Piccirillo and VanAtta (1997). In the case of homogeneous turbulent unstratified shear flow, experiments by Tavoularis and Karnik (1989) and direct numerical simulations by Rogers et al. (1989) support eventual exponential growth of the turbulent kinetic energy K. It is found that, in the stratified case, growth/decay is also exponential. An expression for the non-dimensional growth rate γ is obtained by rewriting the transport equation for the turbulent kinetic energy, Eq. (2.1), to obtain:

$$\gamma = \frac{1}{SK}\frac{dK}{dt} = -2b_{13}(1 - \frac{\epsilon}{P} + \frac{B}{P}) \qquad (2.8)$$

Here b_{13} is the 1-3 component of the shear stress anisotropy tensor

$$b_{ij} = \frac{\overline{u_i u_j}}{\overline{u_k u_k}} - \frac{1}{3}\delta_{ij} . \qquad (2.9)$$

Under the assumption that each term on the RHS of equation (2.8) evolves to an asymptotically constant value for large non-dimensional time St, the equation can be integrated to obtain:

$$K = K_0 \exp(\gamma St) \tag{2.10}$$

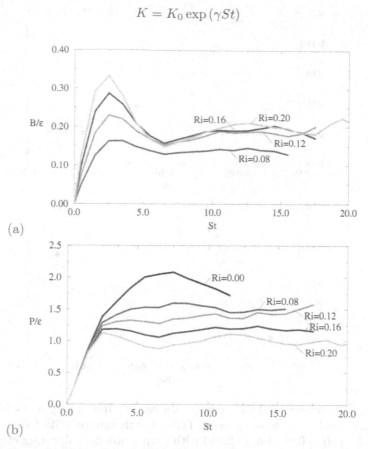

(a)

(b)

Figure 5. Evolution of (a) $-B/\epsilon$ and (b) P/ϵ as a function of the Richardson number Ri. Initial $Re_\lambda = 44.72$ and initial shear number $SK/\epsilon = 2.0$. Reproduced with permission from Jacobitz et al. (1997).

The exponential approximation is also shown with dashed lines in Figure 3a. The constant of integration is used to fit the graphs. The agreement shows that exponential growth or decay of the turbulent kinetic energy K is a good approximation. The exponential decay of K for the case $Ri > Ri_{cr}$ is different from the power law decay observed in unsheared decaying isotropic turbulence behind a grid. Figure 3b shows the evolution of the growth rate γ. Note that a positive value of γ is associated with growth and a negative value with decay of the turbulent kinetic energy K. The asymptotic value of the growth rate γ is positive for small Richardson numbers Ri and negative for large Richardson numbers Ri. The evolution of the anisotropy b_{13} is shown in Figure 4. The magnitude of b_{13} decreases with increasing Richardson number Ri. Therefore the effect of stratification reduces the anisotropy of the flow introduced by the effect of shear. This is a key effect of buoyancy.

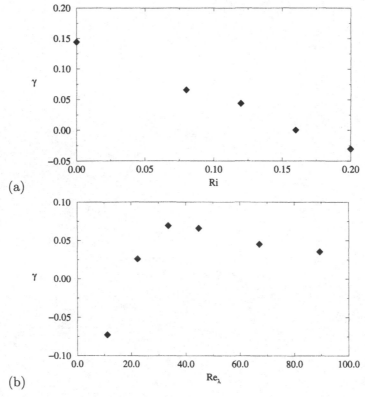

Figure 6. (a) Dependence of TKE growth rate on Ri. Initial $Re_\lambda = 44.72$ and initial $SK/\epsilon = 2.0$ are fixed. (b) Dependence of TKE growth rate on initial Re_λ. The initial $SK/\epsilon = 2.0$ while $Ri = 0.08$. Reproduced with permission from Jacobitz et al. (1997).

In order to identify which term is responsible for stabilization of turbulence, the terms on the r.h.s of Eq. 2.1 are plotted. Figures 5a and 5b show the evolution of $-B/\epsilon$ and P/ϵ, respectively. As expected $-B/\epsilon$ grows with the Richardson number. But this increase is too small to account for the change in the evolution of K. On the other hand P/ϵ decreases strongly with increasing Richardson number. Therefore the primary effect of buoyancy is not a direct sink for K through the buoyancy flux but is the indirect reduction of the shear-induced production of K. This result is in agreement with the previous investigations by Holt et al. (1992) and Rohr et al. (1988).

The dependence of the asymptotic values of the exponential growth rate γ on the Richardson number Ri and the initial values of the Reynolds number Re_λ and the shear number SK/ϵ is now summarized. The asymptotic values of the growth rate γ are computed as the average of the right hand side of equation (2.8) for non-dimensional times $St > 8$. Figure 6 shows the dependence of the growth rate γ on the Richardson number. The growth rate γ decreases approximately linearly with increasing Richardson number Ri. The influence of Reynolds number Re_λ is presented in Figure 6 for fixed

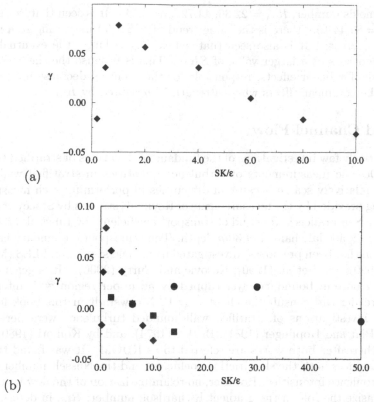

(a)

(b)

Figure 7. (a) Dependence of TKE growth rate on shear number, Sk/ϵ for a low initial $Re_\lambda = 22.36$ series and $Ri = 0.06$. From Jacobitz et al. (1997). (b) Consolidated results for the dependence of TKE growth rate on initial SK/ϵ: low $Re_\lambda = 22.36$ and $Ri = 0.06$ (diamonds), moderate $Re_\lambda = 44.72$ and $Ri = 0.15$ (squares), high $Re_\lambda = 89.44$ and $Ri = 0.15$ (circles). Reproduced with permission from Jacobitz and Sarkar (1999a).

values of initial $Ri = 0.08$ and initial $SK/\epsilon = 2.0$. For low Reynolds numbers, the growth rate γ increases with increasing Re_λ. The growth rate γ reaches a maximum and decreases slightly. For high Reynolds numbers, the growth rate γ tends to become independent of Re_λ. The absence of sensitivity to Re_λ is important because it allows DNS results to be used for verifying and improving turbulence models to be used in high-Re applications. The dependence of the growth rate γ on the initial value of the shear number SK/ϵ is shown in Figure 7a for a series with low initial $Re_\lambda = 22.36$. For low shear numbers, the growth rate γ increases with increasing SK/ϵ, an expected transition from unforced, decaying turbulence to shear-forced, growing turbulence that is also observed in the experiments of Piccirillo and VanAtta (1997). However, the growth rate γ reaches a maximum and decreases with a further increase of SK/ϵ. Note that only a variation of the shear number SK/ϵ can lead to two critical cases with $\gamma = 0$. Figure 7b is a plot that consolidates the dependence of γ on shear number for three different initial choices of Reynolds number, $Re_\lambda = 22.36$, 44.72, and 89.44. It is seen that, for the higher value of $Re_\lambda = 89.44$ too, there is the same trend of $\gamma(SK/\epsilon)$ increasing to a maximum followed by a decrease. It is also seen that, when Re_λ is large, the eventual decrease of growth rate occurs at a larger value of SK/ϵ. This is because the shear needs to be sufficiently large for linear effects, responsible for the eventual decrease in growth rate, to dominate the nonlinear effects whose strength is measured by Re_λ.

3 Stratified Channel Flow

There are relatively few investigations of the fundamental features of stratified turbulence near walls. Oceanic measurements of turbulence quantities in stratified wall-bounded situations are relatively scarce because of difficulties in performing such measurements; examples being provided by the measurements of Reynolds stresses by Stacey et al. (1999) in the northern San Francisco Bay and of transport coefficients by Lu et al. (2000) in the Cordova Channel, a tidal channel of $30m$ depth. The atmospheric boundary layer under stable conditions has been previously investigated using field studies and LES (Mason and Derbyshire (1990); Saiki et al. (1990); Kosovic and Curry (2000)). It is generally found that there is a bottom boundary layer capped by an upper region with internal wave activity; the reader can consult the chapter by F. Nieuwstadt in this book for details. Experimental investigations of stratified wall-bounded turbulence were performed by Arya (1975); Piat and Hopfinger (1981); Britter (1974), and by Komori (1980); Komori et al. (1983) (hereafter both works are referred to as KUOM). It was found that stable stratification reduces both the skin friction coefficient and the Nusselt number as well as suppresses turbulence intensities. However, no relaminarization of the flow was observed. KUOM emphasize the role of the gradient Richardson number, Ri_g, in determining the state of fluctuating motion. Garg et al. (2000) performed LES and DNS of a stratified, incompressible turbulent channel flow at a constant pressure gradient (corresponding to $Re_\tau = 180$) and with various levels of stratification. These authors emphasize the importance of the friction Richardson number, Ri_τ, defined later by Eq. (3.4). Garg et al. (2000) also observed that turbulence was completely suppressed when a moderate stratification, $Ri_\tau \simeq 45$, was exceeded. This result is associated with the initial transient, and when simulations of stratified channel flow are continued on for long time in the LES of Armenio and Sarkar (2002), turbulence is found to be regenerated at the walls after the initial suppression. The linear analysis of Gage and Reid (1968) can be used to show that, for the case with $Re_\tau = 180$ considered by Garg et al. (2000), the corresponding laminar flow would be linearly unstable for $Ri_\tau < 881$, which also implies that the final state at $Ri_\tau = 45 < 180$ cannot correspond to flow relaminarization. Armenio and Sarkar (2002) find that the vertical density gradient sharpens in the outer layer, and can support internal waves which coexist with the inner layer where turbulence, although with reduced transport properties, survives. Correlation coefficients for the the shear stress and the buoyancy flux are found to be primarily determined by Ri_g with an additional weak dependence on Ri_τ.

We will discuss the results of the LES study by Armenio and Sarkar (2002), performed with the dynamic mixed model, in more detail below; the original reference can be consulted for the numerical method and subgrid model. The results of Taylor et al. (2005)

obtained using an adiabatic bottom wall, more akin to the oceanic bottom boundary layer, will also be briefly discussed in order to illustrate the importance of the bottom boundary condition in determining the flow response to stratification.

Table 1. Important quantities that characterize the state of stratified channel flow. Ri_τ is the friction Richardson number imposed on the flow at $t = 0$ which is also the expected theoretical value at the final steady state. Subsequent columns correspond to the final steady state. Reproduced with permission from Armenio and Sarkar (2002).

Case	Ri_τ	Ri_b	Re_τ	Re_b	Re_c	$c_f \times 10^3$	Nu
$C0$	0	0.0	180	2800	3240	8.18	5.95
$C1$	18	0.064	180	3102	3764	6.73	3.71
$C2$	60	0.137	183	3666	4743	4.99	2.79
$C3$	120	0.225	180	4154	5623	3.71	2.27
$C4$	240	0.377	181	4567	6224	3.14	1.70
$C5$	480	0.593	178	5120	7149	2.40	1.28

3.1 Richardson number definitions

In the uniform shear flow studied in section 2, the value of Ri_g applies to the whole flow and is a single parameter that can be used to measure the strength of stratification relative to shear. The situation is more complex in flows with spatial inhomogeneity. It is well known that wall-bounded turbulent flows have two distinct regions: an inner near-wall layer and an outer layer. Outer-scale quantities are centerline mean velocity, u_c, bulk velocity, u_b, and the imposed density difference, $\Delta\rho$. The bulk velocity, $u_b = (1/A) \int u \, dA$, is an area-averaged measure of the channel velocity. The inner-scale velocity is the friction velocity,

$$u_\tau = \sqrt{\frac{\tau_w}{\rho}},$$

whereas the corresponding scale for the density is the friction density,

$$\rho_\tau = \frac{Q_w}{u_\tau} = \frac{-k\frac{\partial \langle \rho \rangle}{\partial z}\Big|_w}{u_\tau}, \tag{3.1}$$

where Q_w is the density flux at the wall and $k = \nu/Pr$ is the molecular diffusivity of density. The friction density is a normalized wall density flux just as the friction velocity is a normalized wall stress. An important nondimensional number which will be referred to throughout the paper is the Nusselt number:

$$Nu = \frac{2hQ_w^{av}}{k\Delta\rho} \tag{3.2}$$

with $Q_w^{av} = (1/2)(Q_{top} + Q_{bot})$. In the case of purely diffusive mass transport through a stationary fluid, $Nu = 1$; thus the value of Nu quantifies the increase of wall mass transport due to turbulence with respect to its laminar value.

It is tempting to parameterize the flow using a single, overall Richardson number. One possibility is to use the outer velocity scale, for example the bulk Richardson number (KUOM),

$$Ri_b = \frac{\Delta \rho g h}{\rho_0 u_b^2} \qquad (3.3)$$

where u_b is the bulk (average) velocity based on flow rate and cross-sectional area. The choice of u_b could be replaced by the free-stream velocity, u_∞, in a developing boundary layer as done by Arya (1975) or by the centerline velocity, u_c, in channel flow as done by Gage and Reid (1968). Another possibility is to use an inner velocity scale and define a friction Richardson number employed by Garg et al. (2000),

$$Ri_\tau = \frac{\Delta \rho g h}{\rho_0 u_\tau^2} \qquad (3.4)$$

The definition of Ri_τ involves an outer-scale quantity for the density variation but an inner-scale quantity for the velocity variation. The relationship between the friction and bulk Richardson numbers is as follows,

$$Ri_\tau = Ri_b \frac{u_b^2}{u_\tau^2} \qquad (3.5)$$

The friction Richardson number is much larger in magnitude than the bulk Richardson number and explicitly involves viscosity.

In unbounded shear flows, the gradient Richardson number, $Ri_g = N^2/S^2$, has been established as the key parameter, based on mean quantities, that determines the influence of stratification. We now estimate values of Ri_g at key locations. The shear at the wall is u_τ^2/ν while, from Eq. (3.2), the mean density gradient at the wall is $(Nu\Delta\rho)/(2h)$ giving the wall-value of the gradient Richardson number as,

$$Ri_g(z=0) = Ri_\tau \frac{Nu}{2Re_\tau^2} \qquad (3.6)$$

which is a small quantity, at high Reynolds number, because of the inverse-square dependence on Re_τ. The variation of Ri_g across the channel can be estimated after approximating the mean velocity and mean density profiles by classical log-layer scaling; the shear is $u_\tau/\kappa z$ and the mean density gradient is $\rho_\tau/\kappa_\rho z$ with κ the Von Karman constant and κ_ρ the equivalent constant for the mean density profile. After using Eqs. (3.1)-(3.2), and some algebraic manipulation the following result is obtained:

$$Ri_g(z) = Ri_\tau \frac{Nu\kappa^2}{2Pr Re_\tau \kappa_\rho} \frac{z}{h} \qquad (3.7)$$

Thus log-layer scaling implies that $Ri_g(z)$ increases linearly in the vertical direction and that, in the core of the channel where $z = O(h)$, the value of gradient Richardson number exceeds the wall value, $Ri_g(0)$ given by Eq. (3.6), by a large multiple, $O(Re_\tau)$. The gradient Richardson number becomes singular at the center of the channel or at the edge of the boundary layer where the mean shear vanishes.

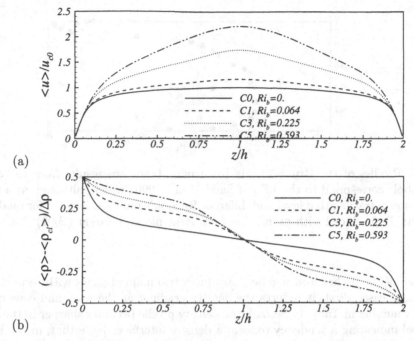

(a)

(b)

Figure 8. (a) Mean velocity, scaled with the centerline velocity of the passive scalar case. (b) Mean density, scaled with the density difference between top and bottom walls. Note that both halves of the channel are shown. Reproduced with permission from Armenio and Sarkar (2002).

3.2 Mean Profiles

Due to the effect of stable stratification, vertical momentum transport is suppressed with respect to the passive scalar case. Since the driving pressure gradient is constant, the viscous wall stress and, therefore, the slope of the mean velocity profile at the wall is invariant between cases as shown in Figure 8(a).

The reduced vertical momentum transport with increasing stratification is manifested as the change in the mean velocity in the core region from a flat, well-mixed profile to a curved profile which, in the limit of relaminarization, would be parabolic. The centerline velocity increases systematically with stratification in Figure 8 and so does the centerline Reynolds number and bulk Reynolds number in Table 1. The friction coefficient, c_f, and the Nusselt number, Nu, are also shown in Table 1. In accord with previous investigations by KUOM and Garg et al. (2000), stable stratification leads to a systematic reduction in both Nusselt number and friction coefficient. Since the simulations are performed with constant mean streamwise pressure gradient (therefore, constant friction velocity at steady state) the reduction of the friction coefficient is due to the increased bulk velocity of the flow.

The mean density profile is also affected by stratification, Figure 8(b). In particular,

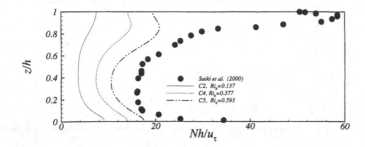

Figure 9. Profiles of the Brunt-Vaisala frequency. Lines are results from the current LES. Symbols correspond to the LES of Saiki et al. (1990) which, although at a significantly stronger stratification level and different Reynolds number, is given for qualitative comparison. Reproduced with permission from Armenio and Sarkar (2002).

the increase of the Richardson number introduces two main changes with respect to the passive scalar case. First, it reduces the density gradient at the wall, and consequently the Nusselt number in Table 1. Second, the density profile becomes sharper in the core of the channel indicating a tendency to form a density interface. Note that, in the limit of complete relaminarization, the mean density variation would be a straight line between the fixed values of the density at $z/h = 0$ (bottom wall) and $z/h = 2$ (top wall). Although, the slope of the density profile at the wall decreases with increasing bulk stratification and tends to that of the laminar profile, the slope at the core of the channel increases and does *not* tend towards the laminar value. For cases C1-C5 the magnitude of the mean density gradient initially decreases away from the wall as expected of turbulent mixing and then, *increases* when the core of the channel is approached.

It is of interest to compare our LES results with those of recent studies of the stable atmospheric boundary layer. Saiki et al. (1990) studied the response of a neutral boundary layer to a cooling heat flux gradually applied to the surface. Their Figure 9 shows a steady state profile of N which, after normalization, is shown along with profiles from the current LES cases in Figure 9. According to our LES results, with increasing stable stratification, N tends to develop a non-monotone behavior, with a local maximum near the wall and another local maximum in the vicinity of the centerline. The results of Saiki et al. (1990), which correspond to a much larger stratification and Reynolds number than our cases, shows an even stronger non-monotone behavior. Internal waves developing in the region between the top of the boundary layer (where the Reynolds shear stress goes to zero and $Ri_g \simeq 0.2$) and the top of the capping inversion are observed by Saiki et al. (1990). In the strongly stratified cases C4 and C5, we also find internal wave activity in the region between $Ri_g \simeq 0.2$ and the center of the channel. Kosovic and Curry (2000) simulated the stable atmospheric boundary-layer for a number of different conditions. All cases showed a turbulent boundary layer above which a region with internal wave activity was observed. To summarize, although the stably-stratified atmospheric boundary layer has different boundary conditions with respect to stably-stratified channel flow,

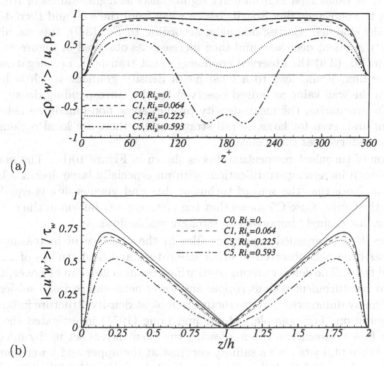

(a)

(b)

Figure 10. The influence of stratification on (a) buoyancy flux, and (b) momentum flux. The plotted flux is the sum of resolved and subgrid model components. Reproduced with permission from Armenio and Sarkar (2002).

both exhibit features generic to stably-stratified wall-bounded flows.

The non-monotone behavior of the mean density gradient (equivalently N) can be explained by considering the Reynolds-averaged density equation which, in channel flow, reduces to the following,

$$< \rho'w' > (z) - k\frac{\partial\rho(z)}{\partial z} = constant \qquad (3.8)$$

where the constant is the wall value of the molecular flux, $-k(\partial\rho/\partial z)_w$. Using the friction density, ρ_τ defined by Eq. (3.1), we rewrite Eq. 3.8 as follows,

$$-\frac{\partial\rho}{\partial z} = -(\frac{\partial\rho}{\partial z})_w (1 - \frac{< \rho'w' >}{\rho_\tau u_\tau}). \qquad (3.9)$$

The last term in Eq. (3.9) is the buoyancy flux (henceforth used interchangeably with the mass flux) normalized using inner variables, and its variation is shown in Figure 10(a). In the passive scalar case, $Ri_b = 0$, the normalized buoyancy flux is almost unity away from the wall, $z/h > 0.1$; this corresponds to the 'constant flux' hypothesis of neutral boundary layers. However, its value departs from unity significantly at higher values of Ri_b. In the stratified cases, the buoyancy flux first increases away from the wall and then decreases in the core of the channel. Correspondingly, according to Eq. (3.9), the magnitude of the mean density gradient decreases and then increases as observed in Figure 8. In case C5, according to Eq. (3.9) the observed countergradient transport, i.e. negative values of the buoyancy flux, would lead to a local mean density gradient which is larger in magnitude than the wall value as indeed observed for the corresponding density profile in Figure 8. To summarize, the mean density profiles show non-monotone behavior of density gradient and, even for large overall stratification, there are local regions which are mixed by boundary-layer turbulence.

The variation of turbulent momentum flux is shown in Figure 10(b). The maximum value decreases with increasing stratification, with an especially large decrease between case C3 and C5. Note that the sum of turbulent flux and viscous flux is equal to the straight line in the figure. Case C5 shows that the viscous contribution in the core of the channel becomes increasingly important in strongly-stable flows.

The LES result of sharpened density profiles in the presence of increasing overall stratification has been also observed in other laboratory and field studies of mixing in stably stratified flows. The density mixing in stratified channel flow can be conceptualized as that between two turbulent regions (upper and lower near-wall regions) subject to an overall stable density difference. It is instructive to look at density structure in flows with a similar configuration. For example, Moore and Long (1971) investigated the stably-stratified shear flow between two counterflowing turbulent layers set up by injection of approximately horizontal jets with a salinity contrast at the upper and lower boundaries of a tank. The authors find that there is a *steady state* vertical density structure that consists of upper and lower regions, almost uniformly mixed by turbulence, separated by a central region where most of the density variation occurs. Crapper and Linden (1974) studied the central region between two layers stirred by a horizontal grid and found that, with increasing stable stratification, the density profile sharpens in the central region and, under some circumstances, a steady density interface is possible. It appears that the sharpening of the density profiles observed in our channel flow LES is consistent with other observations in similar stratified mixing situations.

Velocity profiles, plotted using inner scaling, are also of interest. In the unstratified case, three zones can be identified: the inner region ($z^+ < 30$), the logarithmic region ($30 < z^+ < 150$) where the mean velocity variation is approximately linear in log-linear coordinates, and a small core region for $150 < z^+ < 180$ where there is a deviation from the log law. The classical expression for the law of the wall in the neutral case is:

$$u^+ = A \ln z^+ + B.$$

Stable stratification increases the slope, A, of the log-zone and reduces the intercept, B, of the velocity profile. Also, the core region appears to be widened and characterized by a larger mean velocity gradient as compared to the passive scalar case. The density profiles plotted using wall units show that, similar to the mean velocity profile, there are three zones. Under strongly-stable stratification, the vertical extent of the log-zone is progressively reduced while the core region widens.

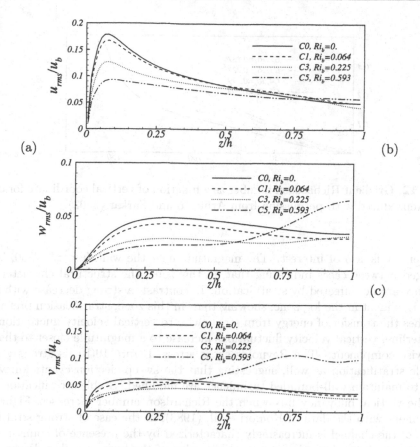

(a)   (b)

(c)

Figure 11. Turbulence statistics in global units for several levels of stratification: (a) Streamwise velocity, (b) Vertical velocity, and (c) Spanwise velocity. Reproduced with permission from Armenio and Sarkar (2002).

3.3 Turbulence Profiles

The streamwise intensity is reduced in most of the channel with respect to the passive scalar case, Figure 11(a). The wall-normal turbulent intensity shown in Figure 11(b) behaves differently depending on the level of stratification. For weakly stratified flow (Cases 1, 2 and 3) it is reduced throughout the channel. In the other cases, the reduction of vertical fluctuations in the wall layer and log-zone with increasing Ri continues; however, the fluctuations are enhanced in the core region. When stratification increases, counter-gradient density fluxes are observed in the core region, Figure 10(a), that act as a source term in the $<w'w'>$ transport equation. The spanwise velocity fluctuations shown in Figure 11(c) decrease with increasing stratification except for case C5. The density fluctuations in the log-zone decrease consistently with Ri_b. However, ρ_{rms} in the core of the channel increases with increasing Ri_b. The ratio between the vertical and the stream-

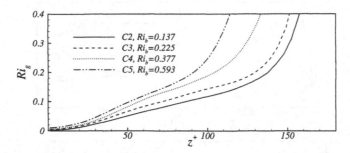

Figure 12. Gradient Richardson number as a function of vertical coordinate for different cases. Reproduced with permission from Armenio and Sarkar (2002).

wise intensity is also of interest. The magnitude near the wall, $0 < z^+ < 30$, is nearly unchanged between cases indicating that, in this zone, the structural characteristics of turbulence are not affected by stratification. In contrast, a strong decrease with increasing Ri_g is evident in the log zone, showing that, in this zone, stratification preferentially diminishes the transfer of energy from streamwise to vertical velocity fluctuations. Near the centerline, vertical velocity fluctuations increase to a magnitude closer to that of the streamwise component. The Reynolds shear stress, Figure 10(b), is strongly reduced by stable stratification as well, suggesting that the sweep-ejection events known to be critical to maintain wall-bounded turbulence are modified. Stable stratification also reduces the vertical buoyancy flux when the Richardson number increases, Figure 10(a). In agreement with the data of Komori et al. (1983), in the case of strong stratification, the core of the channel is increasingly characterized by the presence of counter-gradient density fluxes. Note that, the unnormalized value of the buoyancy flux decreases more dramatically than that suggested by Figure 10(a) because ρ_τ decreases with increasing Ri_b. Similar to the turbulent intensities, the ratio of the vertical and streamwise density fluxes (not shown) is mainly affected away from the wall. The vertical buoyancy flux is preferentially damped relative to the streamwise flux.

3.4 Gradient Richardson number

Profiles of the local value of the gradient Richardson number, $Ri_g = N^2/S^2$, are shown in Figure 12. The gradient Richardson number is singular at the centerline and is very large in a narrow region near the centerline. On the other hand, in a large region from the wall onwards, Ri_g ranges between 0.02 and 0.2 and is well-behaved. There is an equilibrium region where the variation of Ri_g is linear as suggested by Eq. (3.7). For a given case, Ri_g exhibits monotone increase from the wall towards the center of the channel. Also, cases with increasing Ri_τ have larger values of Ri_g at the same distance from the wall. It is remarkable that the value $Ri_g \simeq 0.2$ seems to be the value that determines the point for the abrupt change of the slope of the $Ri_g(z)$ profile.

The ratio $-B/P = Ri_f$, the flux Richardson number, or sometimes, $-B/\epsilon$, is often used to represent mixing efficiency. Here, $B = -\frac{g}{\rho_0} < \rho'w' >$ is the buoyancy flux while

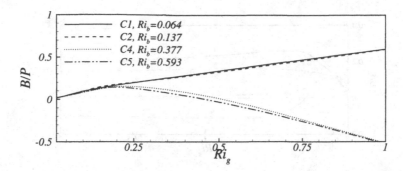

Figure 13. The influence of gradient Richardson number on the flux Richardson number, $Ri_f = -B/P$. Reproduced with permission from Armenio and Sarkar (2002).

$P = -\frac{\partial <\bar{u}>}{\partial z} < u'w' >$ is the production of turbulent kinetic energy. It is worth emphasizing that B represents reversible energy transfer and does not necessarily indicate molecular mixing of the density field. A more direct estimate of molecular mixing would involve measurement of the scalar (density) dissipation and is commonly used in combustion and heat transfer applications. Quantities such as the available potential energy and background potential energy are being increasingly used in the stratified flow literature to distinguish between true (molecular) mixing and buoyancy flux, for example Winters et al. (1995); Winters and DAsaro (1996). Note that, within the framework of the eddy viscosity hypothesis,

$$< u'w' > = -D_u \frac{\partial < u >}{\partial z} \tag{3.10}$$

$$< \rho'w' > = -D_\rho \frac{\partial < \rho >}{\partial z} \tag{3.11}$$

the ratio,

$$\frac{-B/P}{Ri_g} = \frac{D_\rho}{D_u} = \frac{1}{Pr_t} \tag{3.12}$$

where Pr_t is the turbulent Prandtl number. The behavior of the ratio $-B/P$ is worthy of describing in some detail (Figure 13). For $0 < Ri_g < 0.2$, the value of B/P increases linearly with the gradient Richardson number in all the cases investigated. Correspondingly, the turbulent Prandtl number is nearly equal to one in the neutral case and it slightly increases with increasing Ri_g. The value of $Pr_t \simeq 1$ observed here is consistent with previous studies of neutral flows where values of turbulent Prandtl number in the range of $0.7 < Pr_t < 1.2$ have been measured depending on the specifics of the flow. Furthermore, previous laboratory and numerical studies of stratified uniform shear flow, Gerz et al. (1989); Jacobitz et al. (1997), show that Pr_t increases little in the range $0 < Ri_g < 0.2$ just as in the current study. For $Ri_g > 0.2$ the behavior is qualitatively different, depending on the level of stratification. For weak stratification, cases C1 and C2, the value of B/P continues to increase with Ri_g but at a smaller rate and, thus,

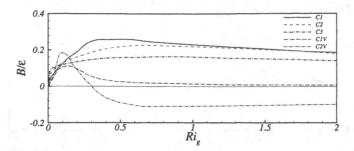

Figure 14. The ratio, B/ϵ, as a function of the gradient Richardson number. C1V and C2V are cases of channel flow with vertical shear at $Re_\tau = 390$, higher than the $Re_\tau = 180$ cases simulated by Armenio and Sarkar (2002). C1, C2 and C3 are cases of channel flow past side walls where the shear is horizontal. Reproduced with permission from Armenio and Sarkar (2004).

the value of Pr_t increases somewhat with increasing Ri_g. For stronger stratification, cases C4 and C5, after attaining a maximum value (approximately 0.14), B/P *decreases* with increasing Ri_g. Consequently, Pr_t increases rapidly. A more rapid increase of Pr_t for $Ri_g > 0.2$ has also been observed in previous laboratory and numerical studies of vertically sheared flow as summarized by Schumann and Gerz (1995), see their Figure 3. The behavior of Pr_t at large values of Ri_g is of interest in developing parameterizations of mixing. If the mixing efficiency, $B/P \to cnst$, then, according to Eq. (3.12), Pr_t would increase linearly with Ri_g. However, reviews of data by Ivey and Imberger (1991) and Schumann and Gerz (1995) suggest that B/P eventually decreases with Ri_g. The strongly stratified cases C4 and C5 show a decrease of B/P when $Ri_g > 0.2$ and, consequently, a dramatic increase of Pr_t which is superlinear. However, cases C1-C2 with weaker overall stratification show a more moderate and quasilinear increase of Pr_t. Finally, in cases C4 and C5, when $Ri_g > 0.45$, countergradient buoyancy flux is observed so that B/P becomes negative. Countergradient buoyancy fluxes have observed previously, for example, by Gerz et al. (1989) and Holt et al. (1992) in uniform shear flow when a 'transition' Richardson number (between 0.4 and 0.7) is exceeded, and by Komori et al. (1983) in the experimental investigation of strongly-stratified channel flow. Such behavior with negative values of B/P is not representative of a turbulent Prandtl number, since the eddy viscosity concept ceases to hold.

The ratio B/ϵ, related to the so-called mixing efficiency, is plotted in Fig. 14 for two cases (CV1 and CV2) of channel flow. The value of B/ϵ decreases sharply at $Ri_g \sim 0.2$ to eventually become almost zero for CV1 ($Ri_b = 0.21$) and becomes eventually negative for CV2 ($Ri_b = 0.36$). Although the values of $Re_\tau = 390$ and $Pr = 5$ in the simulations of Armenio and Sarkar (2004) are higher than the values of $Re_\tau = 180$ and $Pr = 0.71$ employed in the simulations of Armenio and Sarkar (2002), the results of the two studies are similar. The maximum value of $B/\epsilon \simeq 0.2$. The curves labelled C1-C3 in Figure 14 correspond to channel flow past side walls where the mean shear is *horizontal*. The ratio

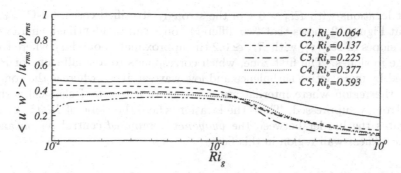

Figure 15. Correlation coefficient of the momentum flux as a function of the gradient Richardson number. Reproduced with permission from Armenio and Sarkar (2002).

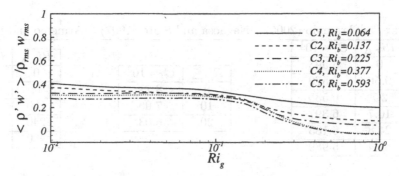

Figure 16. Correlation coefficient of the density flux as a function of the gradient Richardson number. Reproduced with permission from Armenio and Sarkar (2002).

B/ϵ is larger when the shear is horizontal and, furthermore, does not exhibit the sharp decrease at large Ri_g that is observed in cases CV1-CV2.

As discussed in the introduction, the local value of the gradient Richardson number, Ri_g, is generally recognized as the most important *local* parameter for measuring stratification in shear flows without walls and to interpret field measurements of ocean microstructure. In the present flow, although walls are present, the importance of Ri_g is again evident. Figure 15 shows the correlation coefficients associated with the momentum flux for all the simulated cases. Although there is also a systematic decrease with Ri_τ, the different curves exhibit the same behavior as a function of Ri_g. After a relatively constant value at low values of the gradient Richardson number, the large decrease in the value of the correlation coefficient over a narrow range, $0.15 < Ri_g < 0.2$, is remarkable. This decrease in the correlation coefficient indicates a rapid change from unstratified turbulence with classical mixing characteristics to stably-stratified turbulence with inhibited vertical mixing. The correlation coefficient associated with the buoyancy flux in Figure 16 also shows similar features. In addition, countergradient buoyancy fluxes are present at locations with $Ri_g > 0.4$ in the strongly stratified cases, C4-C5. Finally a closer look at Figure 12 shows that two different zones can be identified, characterized with different slopes of $Ri_g(z)$, with $Ri_g \simeq 0.2$ the approximate boundary where there is a sharp change in slopes. The first zone, which corresponds to a small slope of $Ri_g(z)$, is characterized by more active turbulence and low wave-activity, whereas the zone with large slope is the region where internal waves are more active and classical turbulent transport is strongly suppressed. Thus, the location where the slope of $Ri_g(z)$ increases rapidly, indicates the border between the *buoyancy-dominated* central region and the *buoyancy-affected* near-wall region in this flow.

Table 2. Friction Coefficient: Taylor, Sarkar and Armenio (2005) has an imposed surface heat flux at the upper surface and adiabatic lower wall. Nagaosa and Saito (1997) have an upper free surface and lower wall, both being isothermal. Armenio and Sarkar (20002) have upper and lower walls, both being isothermal.

Taylor, Sarkar and Armenio (2005)

Ri_τ	$Ri_{\tau,\Delta}$	$C_f * 10^3$
0	0	6.593
25	0.56	6.579
100	2.7	6.535
250	10.7	6.397
400	24.8	6.183
500	39.4	5.989

Nagaosa and Saito (1997)

$Ri_{\tau,\Delta}$	$C_f * 10^3$
0	8.71
10	7.06
20	6.03

Armenio and Sarkar (2002)

$Ri_{\tau,\Delta}$	$C_f * 10^3$
0	8.18
18	6.37
120	3.71
240	3.19
480	2.40

3.5   Case with Adiabatic Bottom Wall

Since all previous studies of channel flow have considered fixed temperature walls, the near-wall region was stratified leading to a reduction of the near-wall turbulence production. When considering environmental flows, the results of these studies may be analogous to the atmospheric surface boundary layer under conditions of strong surface cooling where a stably stratifying heat flux at the ground can lower turbulent production in the surface layer, Mahrt (1999). On the other hand, in the oceanic boundary layer the bottom surface is adiabatic, see Lien and Sanford (2004) for a good explanation of the differences between atmospheric and oceanic boundary layers. The oceanic bottom boundary layer has motivate Taylor et al. (2005) to examine channel flow with an adiabatic bottom wall. The upper surface is a free surface (assumed plane) with an imposed heat flux that is increased between cases to increase stratification effects.

Table 2 gives the skin friction coefficient, C_f, for each case of Ri_τ. For comparison, the values found by Nagaosa and Saito (1997) and Armenio and Sarkar (2002) are also shown. $Ri_{\tau,\Delta}$ defined with the density difference across the channel,

$$Ri_{\tau,\Delta} = \frac{gh\Delta\rho}{\rho_0 u_\tau^2}, \tag{3.13}$$

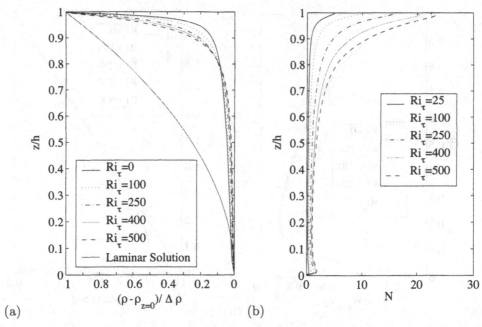

Figure 17. Stratified channel flow with adiabatic bottom wall: (a) Mean density profile, and (b) Profile of buoyancy frequency. From Taylor et al. (2005).

is introduced to measure stratification on a similar basis in all studies. Clearly C_f decreases with $Ri_{\tau,\Delta}$ in all studies, but the dependence observed here is much *weaker* than the 31% decrease between $Ri_{\tau,\Delta} = 0$ and 20 observed by Nagaosa and Saito (1997) and the 22% decrease between $Ri_{\tau,\Delta} = 0$ and 18 observed by Armenio and Sarkar (2002). This can be explained by the relatively limited region affected by stratification in the present study, a qualitative difference with respect to the previous cases with fixed temperature boundary conditions. The mean density profile for each case is plotted as a function of nondimensional height in Figure 17(a) where the density is made nondimensional by $\Delta\rho$, the difference between wall and surface values as in Komori et al. (1983). The laminar solution is also shown. Unlike the gradual variation of $\rho(z)$ in the laminar case, the turbulent flow exhibits a strongly stratified region, or pycnocline, near the free surface that overlies a well-mixed region near the lower wall. The presence of the well-mixed region must depend on the existence of active turbulence since the density gradient of the laminar solution vanishes only near the wall. The thickness of the pycnocline increases with Ri_τ, implying that the turbulence generated near the lower wall is less effective at mixing for large Ri_τ. It should be noted that the density gradient is small but nonzero and nearly constant in the lower portion of the channel and only vanishes in a very thin layer within about 5 wall units from the wall. Figure 17(b) shows the buoyancy or Brunt-Vaisala frequency, Nh/u_τ. This plot also makes clear the deepening and strengthening of the pycnocline with increasing Ri_τ. Comparison of the profiles

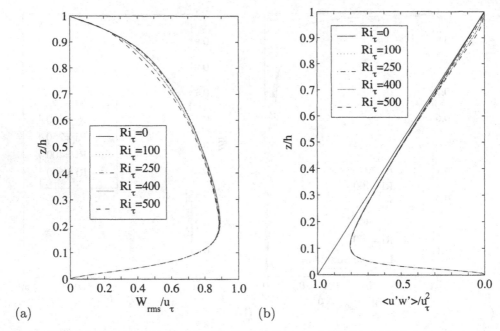

Figure 18. (a) R.m.s vertical velocity fluctuation, and (b) Reynolds shear stress From Taylor et al. (2005).

of N in Figure 17(b) with those in Figure. 9 shows that the lower water column has substantially less stratification in the case with adiabatic bottom wall which, as will be seen, lead to important differences in turbulence characteristics with respect to stratified situations with constant temperature or imposed heat flux at the bottom.

Figure 18(a) shows the profile of the r.m.s. vertical velocity, w. In the lower half of the channel, the profiles collapse and are similar to that in unstratified closed channel flow. In the upper region, w_{rms} decreases monotonically with increasing Ri_τ. Since w_{rms} corresponds to the vertical turbulent kinetic energy, and Ri_τ is linked to the size of the buoyancy suppression term in the TKE budget, the observed decrease is as anticipated. Interestingly, near the free surface where w_{rms} is supressed by the geometry, the dependence on Ri_τ is lost. The Reynolds shear stress, $< u'w' >$ shown in Figure 18(b) along with the total shear stress, $\tau(z) = \tau_{wall}(1 - z/h)$. The viscous shear stress is the difference between this line and $< u'w' >$. Stratification induces a decrease in the magnitude of $< u'w' >$, which is especially strong when $Ri_\tau = 500$. Although the influence of Ri_τ on the bulk Reynolds stress is rather small as also reflected in small changes to the friction coefficient, the *local* turbulent diffusion is strongly affected in a significant portion of the channel, as can be seen by examining the eddy viscosity, ν_T (not shown).

We had shown earlier in Figure 15 the behavior of the u-w correlation coefficient in simulations with constant bottom wall temperature. Fig. 19 shows the analogous plot

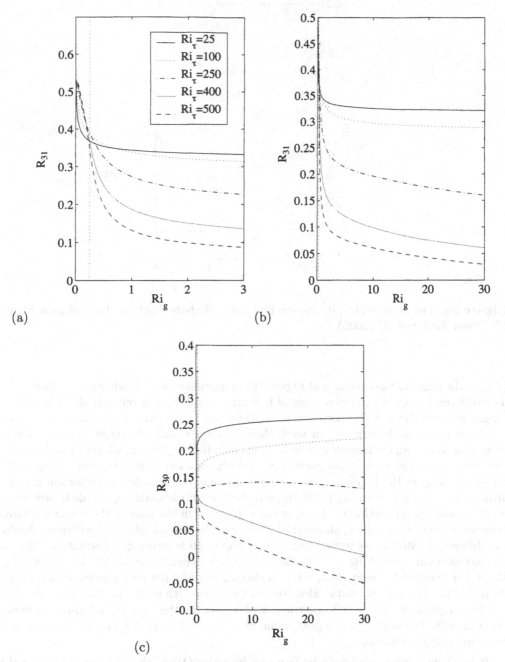

(a)

(b)

(c)

Figure 19. Correlation coefficients as a function of Ri_g: (a) Streamwise and vertical velocity correlation, (b) Same as (a) except for an expanded range of Ri_g, and (c) Density and vertical velocity. The vertical line denotes $Ri_g = 0.25$. From Taylor et al. (2005).

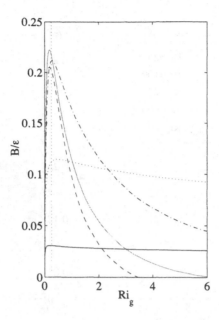

Figure 20. The ratio, $-B/\epsilon$, in channel flow with adiabatic bottom. Legend as in Figure 19. From Taylor et al. (2005).

(R_{13} is the same as the ordinate of Figure 15) in simulations with adiabatic bottom wall. The difference between the two types of boundary conditions is remarkable. There is no longer a sharp drop of the correlation coefficient in the region $0.1 < Ri_g < 0.25$ that was seen earlier in Figure 15. Instead, there is a systematic decrease as a function of increasing Ri_g from the beginning which tapers off when Ri_g becomes large. Furthermore, the decrease of the correlation coefficient between cases is much stronger in Figure 19 relative to Figure 15. Thus, a universal dependence of the correlation coefficient on Ri_g, although useful for developing RANS-type turbulence models and applicable to situations like the atmospheric boundary layer, is not supported in the case of the oceanic bottom boundary layer. Figure 19(c) shows the behavior of the ρ-w correlation coefficient. Again, the difference with the corresponding plot in Figure 16 is striking. The ratio, $-B/\epsilon$, is plotted as a function of Ri_g in Fig. 20. There is a dependence on Ri_τ as well as Ri_g. When the value of Ri_τ is modest, $-B/\epsilon$ is also modest and remains almost constant after its initial rise from a zero value. However, in the cases with high Ri_τ, the value of $-B/\epsilon$ exhibits a peak at $Ri_g \sim 0.25$ (vertical dashed line in the figure) and then decreases substantially. Interestingly, the peak value of $-B/\epsilon$ is close to 0.2, the value often used in oceanographic models.

It has been shown previously by Pan and Banerjee (1995) that a large portion of the Reynolds stress near an unstratified free surface in open channel flow is due to impinging of low-speed fluid advected from the near wall region. While the wall generated low-speed streaks do not maintain coherence over distances comparable to the channel height in

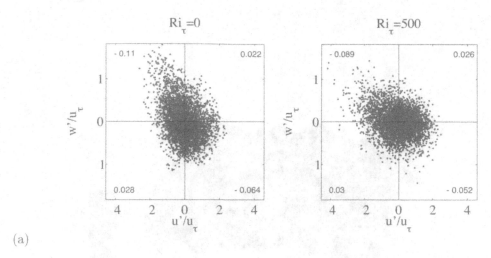

Figure 21. Scatter plot of w' versus u' at $z/h = 0.84$. From Taylor et al. (2005).

this study, low-speed ejections from the wall boundary layer are observed to directly impact the free surface in the low stratification cases. That the upward advection of low speed fluid to the surface is inhibited for large Ri_τ is implied by the drop in correlation between u' and w' in the Reynolds stress of figure 18(b). Contributions to the Reynolds stress can be seen by plotting u' vs. w' as shown in Figure 21 for $z/h = 0.84$.

In each quadrant of the plots is a label showing its contribution to $< u'w' > /u_\tau^2$. The upwelling events can be clearly seen for $Ri_\tau = 0$ by an anisotropic tail extending to the upper left. When $Ri_\tau = 500$ the strength of the upwellings is diminished, and the distribution becomes more isotropic. In both cases, downwelling events are not as energetic as upwelling bursts, and contribute less to $< u'w' >$. The effect of stratification on dense fluid upwellings near the free surface can also be clearly seen by examining the instantaneous property distributions. Figure 22 shows ρ' and w', the deviation from the horizontal mean, at $z/h = 0.999$ for $Ri_\tau = 0$ and $Ri_\tau = 500$ at the last simulation time in both cases. The height of the surface mesh denotes the vertical velocity with the tall peaks indicating rising fluid ($w' > 0$). The corresponding grayscale shows ρ' with dark gray denoting heavy fluid with positive ρ'. Notice that for $Ri_\tau = 0$, each region of upwelling is associated with a positive density anomaly indicating an upwelling of dense fluid from the bottom. When $Ri_\tau = 500$, almost none of the positive w' patches at this particular time are associated with large positive ρ'. These snapshots are typical of those seen throughout the simulation; while the existence of dense fluid upwellings cannot be precluded for the strongest stratification cases, they are much less common than when $Ri_\tau = 0$.

4 Uniform Horizontal Shear and Uniform Stratification

Jacobitz and Sarkar (1998) performed a series of direct numerical simulations, in which

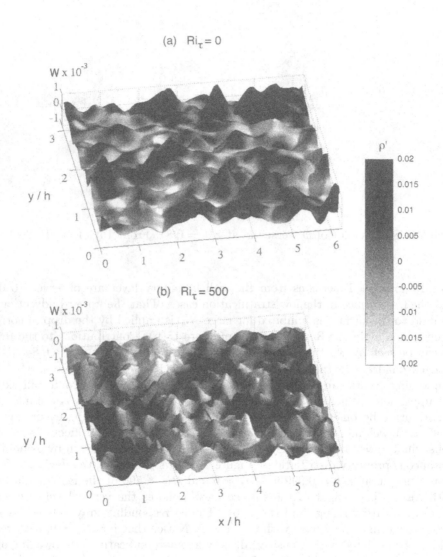

Figure 22. Instantaneous height maps of vertical velocity with density perturbations in grayscale. From Taylor et al. (2005).

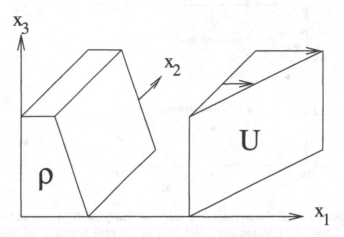

Figure 23. Schematic of uniform horizontal shear flow. Reproduced with permission from Jacobitz and Sarkar (1999b).

the angle θ between the gradient of the mean density and the gradient of the mean stream-wise velocity component was varied continuously from $\theta = 0$, corresponding to the vertical shear case, to $\theta = \pi/2$, corresponding to the horizontal shear case. Let S be the magnitude of the shear given by $\sqrt{S_H{}^2 + S_v{}^2}$ when both horizontal and vertical components are present. Stratification and shear impose time scales on the turbulence evolution. The time scale introduced by the density stratification is $T_\rho = 1/N$, where $N = (-gS_\rho/\rho_0)^{1/2}$ is the Brunt-Väisälä frequency. The time scale introduced by the velocity shear is $T = 1/S$, where S is the shear rate with the unit of a frequency. The generalized gradient Richardson number $Ri_g = N^2/S^2$ is the square of the ratio of the two time scales and therefore the primary parameter describing the relative influence of stratification and shear on the turbulence evolution. Jacobitz and Sarkar (1998) find that there are values of Ri_g for which the turbulence evolution changed from decay for $\theta = 0$ to growth for $\theta = \pi/2$. A physical explanation of the enhanced turbulence levels in the case of horizontal shear is as follows. When the mean shear is horizontal, the sustenance of turbulent velocity fluctuations involves production by *horizontal* eddies interacting with the mean shear $S_H = \partial \overline{U}_1/\partial x_2$. Buoyancy does not directly inhibit the horizontal motion. However, in the vertical shear case, turbulence requires a self-sustaining inter-action between the mean shear $S_V = \partial \overline{U}_1/\partial x_3$ and *vertical* eddies. Gravity directly inhibits these vertical overturns. Hence, for the same values of the mean shear and strat-ification rates, the production is reduced when the mean shear is vertical rather than horizontal. Jacobitz and Sarkar (1999b) focussed on the case of pure horizontal shear, see Figure 23, and reported on turbulence energetics, transport, and anisotropy. We give some highlights from Jacobitz and Sarkar (1999b) below.

It is known that, in stably stratified uniform shear flow, the asymptotic value of the growth rate γ reaches a constant value and turbulent kinetic energy shows exponential

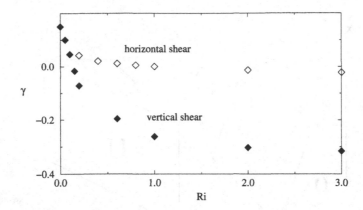

Figure 24. Dependence of the asymptotic value of the growth rate γ on the Richardson number Ri for the vertical (filled symbols) and horizontal (open symbols) shear cases. Reproduced with permission from Jacobitz and Sarkar (1999b).

growth or decay. Figure 24 shows the dependence of the asymptotic value of the growth rate $\gamma = (1/SK)dK/dt$ on the Richardson number Ri for the vertical and horizontal shear cases. The asymptotic value is computed as the average of values in the asymptotic range with $St \geq 8$. The growth rate is positive for small values of Ri, corresponding to growth of K, and it is negative for large values of Ri, corresponding to decay of K. The case $\gamma = 0$ corresponds to a constant level of K and the Richardson number assumes its critical value $Ri = Ri_{cr}$. According to Fig. 24, the value of $Ri_{cr} \simeq 1.5$ for the horizontal shear case is an *order of magnitude* larger than the corresponding value of Ri_{cr} in the vertically sheared case.

Diapycnal transport due to turbulence is now discussed. Initially, a large buoyancy flux is observed that converts kinetic energy into potential energy, because all simulations are started without density fluctuations. The buoyancy flux B decreases as the Richardson number Ri is increased for both vertical and horizontal shear cases. Thus, the stronger stratification suppresses vertical mass transport. However, a comparison of the vertical and horizontal shear cases with $Ri = 0.2$ shows that the horizontal shear results in a stronger buoyancy flux than the vertical shear case.

The buoyancy flux B is usually estimated from the dissipation rate ϵ in geophysical field studies:

$$B = \Gamma \epsilon. \tag{4.1}$$

Figure 25 shows the dependence of the mixing efficiency Γ computed from the direct numerical data on the Richardson number Ri for the vertical and horizontal shear cases. The mixing efficiency Γ is computed as the average in the asymptotic range with $St \geq 8$ for which Γ has reached an approximately constant value. In the vertical shear case, the mixing efficiency reaches the predicted value $\Gamma = 0.2$. However, the weakly stratified cases with $Ri < 0.1$ result in smaller values for Γ that would lead to an over-prediction of B. The strongly stratified case with $Ri = 2.0$ has a counter-gradient buoyancy flux and

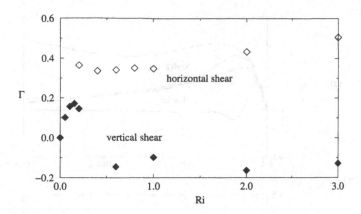

Figure 25. Dependence of $\Gamma = -B/\epsilon$ on the Richardson number Ri for the vertical (filled symbols) and horizontal (open symbols) shear cases. Reproduced with permission from Jacobitz and Sarkar (1999b).

Γ is negative. The observation of a counter-gradient buoyancy flux for $Ri = 2.0$ agrees with the results of Gerz et al. (1989). In the horizontal shear case, the mixing efficiency assumes a higher value $\Gamma \approx 0.4$. The buoyancy flux may therefore be under-predicted by the assumption of $\Gamma = 0.2$ in the presence of horizontal shear.

Turbulent transport of mass and momentum can be studied by consideration of the eddy viscosity ν_t of the velocity field and the eddy diffusivity α_t of the density field. In the vertical shear case, the vertical gradients of mean velocity and density result in a turbulent transport in the vertical direction:

$$\nu_t = -\frac{\overline{u_1 u_3}}{S_V} \qquad \alpha_t = -\frac{\overline{u_3 \rho}}{S_\rho} \tag{4.2}$$

However, in the horizontal shear case, the horizontal velocity gradient results in a turbulent transport of momentum in the horizontal direction. The vertical density gradient again results in a turbulent transport of mass in the vertical direction:

$$\nu_t = -\frac{\overline{u_1 u_2}}{S_H} \qquad \alpha_t = -\frac{\overline{u_3 \rho}}{S_\rho} \tag{4.3}$$

In the following, the turbulent diffusion coefficients are computed from the direct numerical data and the vertical and horizontal shear cases are compared.

Figures 26a and 26b show the evolution of the eddy viscosity ν_t of the velocity field and the eddy diffusivity α_t of the density field, respectively. Cases with moderate (Ri=0.2) and strong (Ri=2.0) stratification as well as with vertical (solid lines) and horizontal (dashed lines) shear are compared.

The $Ri = 2.0$ cases show strongly reduced eddy diffusivities ν_t and α_t compared to the $Ri = 0.2$ cases. However, for constant Richardson number, the eddy diffusivities are an *order of magnitude larger* in the horizontal shear case that in the vertical shear case. That

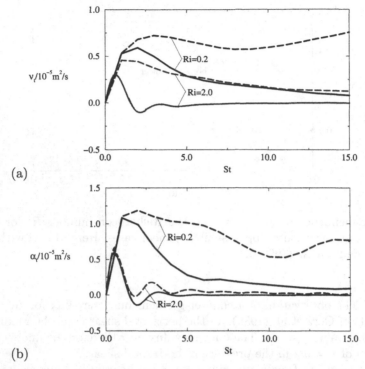

Figure 26. Comparison of (a) the eddy viscosity ν_t and (b) the eddy diffusivity α_t for the vertical (solid lines) and horizontal (dashed lines) shear cases. Reproduced with permission from Jacobitz and Sarkar (1999b).

the *vertical* eddy diffusivity of mass is an order of magnitude larger when the mean shear is *horizontal* is a striking result. Although counter-intuitive at first glance, this result can be understood by recognizing that the horizontal eddies promoted by the mean horizontal shear are *not* directly inhibited by buoyancy, and that the energy transfer between velocity components as well as vortex stretching typical of three-dimensional turbulence is sufficient to induce vertical motion from the strongly energetic horizontal eddies. It was shown earlier that the production is much larger in the horizontal shear case leading to larger turbulent kinetic energy. Since the vertical energy partition $\overline{u_3 u_3}/2K$ of turbulent kinetic energy does not change substantially due to stratification, see Jacobitz and Sarkar (1999b) the vertical velocity fluctuations and the associated vertical mass diffusivity are larger when the mean shear is horizontal as is observed in the direct numerical simulations.

5 Shear Layer with Horizontal Shear

Shear layers with horizontal shear have been observed in the ocean. Flament et al. (2001) have studied a zonal horizontal shear flow off the island of Hawaii using drifting

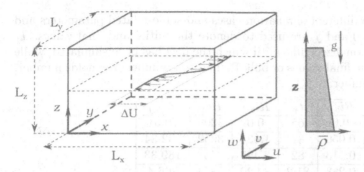

Figure 27. Schematic of a temporally evolving horizontal shear flow with vertical stratification. From Basak and Sarkar (2005).

buoys and sea-surface height anomaly and find that the initially thin shear layer grows by vortex pairing. Munk et al. (2000) show visualizations of sea-spiral eddies obtained using sunglint, and analyze the dynamics of the spiral eddies. A nice example of KH rollers in a horizontal shear layer is provided by their Fig. 2. Horizontal shear instabilities at the Norwegian Coastal Current front were observed using synthetic aperture radar by Johannessen et al. (1996). The scale of the horizontal shear in the preceding examples is of the order of 1 to 100 kilometers and rotational effects are important. Farmer et al. (1995), combining acoustic imaging with CTD surveys, observed strong horizontal shears, $0.02 - 0.10s^{-1}$, on a scale of 100 m in a tidal convergence front in the Haro Strait. It is very difficult to quantitatively observe horizontal shear at scales less than tens of meters although there have been a handful of efforts as discussed by Müller et al. (1986). Turbulent wakes behind spheres and other bluff bodies have both horizontal and vertical shear. The flow instabilities associated with the shear layer velocity profile (summarized later) are different from those in the wake or the jet. A notable difference, immediately apparent, is that the shear layer has Kelvin-Helmholtz (KH) rollers having the same sign of vorticity unlike the shedding of opposite-signed vortices in a wake and the emergence of dipolar vortices in a jet. It should be noted that both, uniform shear flow and channel flow, are not susceptible to inviscid instabilities that will likely play a major role in the shear layer with inflectional mean profile to be studied here. We will now discuss some highlights of the study by Basak and Sarkar (2005) who use DNS to investigate the problem of a shear layer that grows horizontally between two parallel, vertically-infinite fluid streams with a relative horizontal velocity, see Fig. 27, in a stratified medium with constant buoyancy frequency, N.

The different cases simulated and some of the related initial and final quantities are given in Table 3. For all the cases, the size of the computational box is $[L_x \times L_y \times L_z] = 107.5 \times 64.5 \times 32.25$, length being nondimensionalized with the initial vorticity thickness. The number of nodes in the x, y and z directions, $N_x \times N_y \times N_z$ is $640 \times 384 \times 192$ (47 million grid points). The grid spacing is uniform with $\Delta x = \Delta y = \Delta z$, and sufficiently small to resolve the Kolmogorov scale. The same initial velocity field is used in all the simulations, and density fluctuations are initially absent. One simulation

Table 3. Description of the different simulation cases and some related parameters and bulk quantities. Subscripts 0 and f are used to denote the initial and final values. t_f denotes the final nondimensionalized time until which the flow has been simulated, while $Ri_{b,f}$, Re_f, and $Re_{\lambda,f}$ are the final values of bulk Richardson number, Reynolds number, and microscale Reynolds number.

Case	$Ri_{b,0}$	t_f	$Ri_{b,f}$	Re_f	$Re_{\lambda,f}$
A0	0.0	65	0.0	4309	70.51
A1	0.056	82	1.627	3652	132.34
A2	0.113	82	3.457	3763	189.33
A3	0.283	81.2	8.920	3823	266.4
A4	0.567	81.3	21.18	4166	314.42
A5	1.132	79.8	46.00	4341	281.95

without stratification has been done (case A0, $Ri_{b,0} = 0$) to compare the behavior of stratified shear layer with the corresponding unstratified case. The important simulation parameters are the initial Reynolds number Re_0, initial bulk Richardson number $Ri_{b,0}$ and Prandtl number Pr. The bulk Richardson number is defined using the vorticity thickness, δ_ω, of the shear layer as follows,

$$Ri_b = -\frac{g}{\rho_o}\frac{d\bar{\rho}}{dz}\frac{\delta_\omega^2}{\Delta U^2} = \frac{N^2\delta_\omega^2}{\Delta U^2}. \tag{5.1}$$

The values of $Re_0 = 681$ and $Pr = 1$ are kept constant between cases. The value of N^2 is varied keeping ΔU and $\delta_{\omega,0}$ constant, and thus the value of $Ri_{b,0}$ also varies. The temporal shear layer spreads in the cross stream direction with time. The flow statistics vary in the cross stream direction and with time. Quantities such as Re and Ri_b, defined using integrated or maximum (minimum) values, change only with time and serve as indicators of the gross state of the flow. The final Reynolds number, Re_f, given in Table 3 along with the broad-band initial fluctuations is large enough for turbulent flow (Rogers and Moser, 1994). The large values of final Re_λ in the stratified cases is due to the decreased dissipation, ϵ, and the flow field remains well-resolved.

5.1 Evolution of the Mean Flow

Fig. 28(a) shows the evolution of the mean velocity profile, $\bar{u}$, in case A2, and is representative of all the cases simulated. The velocity profile thickens in the cross-stream direction with increasing time. The corresponding growth of the momentum thickness for the different cases is shown in Fig. 28(b). The momentum thickness of the unstratified shear layer grows almost linearly with time. For the stratified cases, the initial growth of the momentum thickness is smaller, however the larger stratification cases (A3-A5) shows an increased growth rate at later times owing to less loss to turbulent dissipation. Since δ_θ increases with time and ΔU is fixed, the mean shear decreases with time, and correspondingly the bulk Richardson number, Ri_b, grows with time reaching greater values for larger initial Ri_b. The late-time growth of the momentum thickness can be

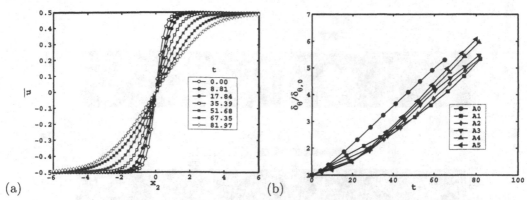

(a)   (b)

Figure 28. (a) Mean velocity profiles at various times. (b) Corresponding values of momentum thickness plotted as a function of time. From Basak and Sarkar (2005).

approximated by the linear law

$$\delta_\theta = c_1 \Delta U t, \qquad (5.2)$$

where the growth rate, c_1, varies between cases. The DNS shows that vorticity thickness is related to the momentum thickness by $\delta_\omega = c_2 \delta_\theta$ with $c_2 \sim 5$, so that $\delta_\omega \sim 5 c_1 \Delta U t$. Using the definition of Ri_b, Eq. (5.1), a quadratic growth law of the bulk Richardson number is found, The strong growth of Ri_b with Nt quickly blurs the differences in its initial value between cases, and results in clustering of the Ri_b profiles at later times. Values of Ri_b greater than 0.5 in cases A1-A5 suggest that stratification effects are likely to become important in all the stratified cases simulated here.

5.2 Structure of the vorticity field

Owing to stratification, the vorticity evolution starting from the same initial conditions has substantial differences between cases. These differences are illustrated by comparing the vorticity field for the unstratified case A0, moderately stratified case A2 and strongly stratified case A5.

Figure 29 shows the vertical (spanwise) vorticity, $|\omega_z|$, at similar times for three cases. The gray-scale scheme in this and following figures is such that black corresponds to the minimum value and white to the maximum. In the unstratified case, Figure 29(a), although lumps of vorticity can be seen, cores and braids representative of the KH (Kelvin-Helmholtz) instability are not distinct. Furthermore, there is little vertical coherence of the vorticity field. The limited spatial organization of the vorticity field is consistent with previous studies of the unstratified shear layer, for example Bell and Mehta (1990) and Rogers and Moser (1994), where the Reynolds number is high and the initial disturbance field is broadband. With increasing stratification, the core/braid structures become increasingly prominent as shown by 29(b)-(c). These *coherent* structures are initially columnar and emerge from the random sea of turbulence, becoming distinct roughly when $Ri_b > 0.5$. The KH rollers that emerge in the stratified shear layer exhibit pairing, tearing and amalgamation, similar to previous observations in high-Re

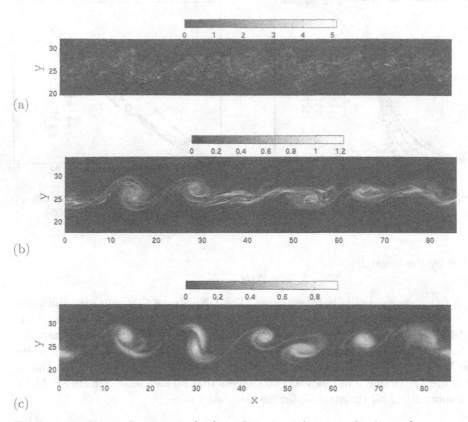

Figure 29. Vertical vorticity, $|\omega_z|$, is shown on the same horizontal cut, $z = 12.7$, and at similar values of the time: (a) Unstratified case A0 at $t = 64.8$, (b) Moderately stratified case A2 at $t = 67.3$, $Ri_b = 2.25$, and (c) Strongly stratified case A5 at $t = 66.1$, $Ri_b = 33.35$. From Basak and Sarkar (2005).

shear layers exhibiting coherent structures, Hernan and Jimenez (1982) and Dimotakis and Brown (1976). The stratification effect that encourages the formation of organized KH rollers in spite of the high Re and disorganization in the initial conditions is an important result of this work. This effect is linked to the inhibition of vertical transport of fluid by stratification which effectively suppresses overturning horizontal vorticity. Besides, the motion of vorticity filaments in vertical planes, causing vertical mass transfer is subdued. These buoyancy effects cause the natural two-dimensional inviscid instability of the free shear layer to become increasing important resulting in the prominence of core dynamics and development of core and braid structures in the flow. The emerging quasi-vertical vortex cores exhibit some streamwise and spanwise variability owing to the phase variation of the small-scale velocity fluctuations in the initial field. At early times, vortex lines in the core are prevented from having large-scale deviations from the vertical, thus constraining the vortex cores to remain quasi-vertical in contrast to the unstratified case.

However, later in time, significant vertical variability develops in the vortex cores as shown by Figure 30. The higher stratification of case A5 results in the formation of clean vortex cores with strong vertical coherence at quite early time, Fig. 30(a). The thin strands of spanwise vorticity noted in the inter-roller zones, are projections of the thin vertical ω_z sheets. These are associated with transfer of spanwise vorticity in the braid regions during pairing of vortex rollers. The random initial condition used in this investigation results in the spanwise and streamwise variability of the ω_z core structures. Fig. 30 sheds light on the spanwise coherence of the interactions of KH rollers that form early on. As seen in Figs. 30(a) and 30(b), these core interactions produce long continuous structures which are inclined to the vertical. One example of this kind of continuous ω_z structure spanning the entire flow domain is marked AA in Fig. 30(a). Visualizations of the corresponding density perturbation field as well as movies of the vorticity and density perturbation fields created with data stored at small time intervals, show that the unstable density field associated with the vortex cores plays a significant role in tilting of the core vortices. Local interactions of these tilted cores with their neighbors modifies the initially continuous vortex core to form a set of *locally* vertical structures. This is exemplified by Fig. 30(b), where the previously continuous structure AA evolves to form a set of stubby vertical structures by sliding at the zones of dislocations. One such dislocation zone is marked by an arrow (B) in this figure. Further examination reveals that these regions of sliding are associated with significant horizontal vorticity. The collapse of the unstable density field creates small vertical scales later in time, Fig. 30(b), from the relatively continuous cores of Fig. 30(a). For example, the truncated structures C and D in Fig. 30(b), that are created from the continuous columns of ω_z at similar locations in Fig. 30(a), further collapse to create thin regions, also marked C and D in Fig. 30(c). There is generation of ω_h in zones C and D that is associated with the collapse of the vertical vorticity. With time, the vertically limited ω_z rollers continue to interact with their neighbors mainly by local pairing. This process leads to further lateral sliding of the rollers resulting in generation of sites of dislocation with significant ω_h, and some ω_z as well. The projection of one of those vortex sheet is marked as B in Fig. 30(c). At late times, the spanwise vorticity forms a lattice of rectangular elements with thick and stubby vertical columns of vertical vorticity and thin horizontal sheets of

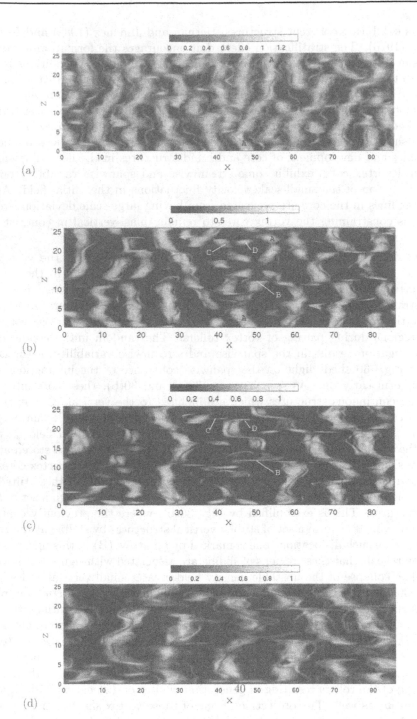

Figure 30. Case A5, spanwise vorticity magnitude, $|\omega_z|$, on a vertical plane. $t = 36.5$, 51.5, 66.1, 79.8. $Ri_b = 2.87$, 8.27, 17.24, 33.35, and 46. From Basak and Sarkar (2005).

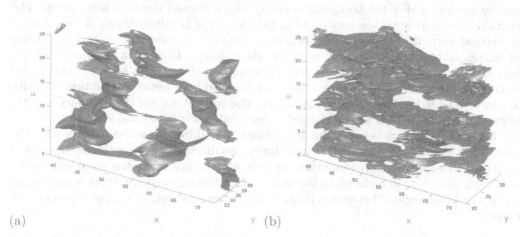

(a)

(b)

Figure 31. Case A5, three dimensional structure of (a) the $|\omega_z|$ field and (b) the ω_h field at time $t = 79.8$. Isosurface value $= 0.4$. From Basak and Sarkar (2005).

predominantly horizontal vorticity. This can be seen clearly in the vorticity isosurface visualization in Fig. 31 as well as in the center 'y' plane visualization of the vorticity field in Figure 30(d). Thus, the late time vorticity field of turbulent horizontal shear flows with high Ri_b, beginning from three-dimensional random velocity perturbations consists mainly of rectangular lattices. This is in contrast to the hexagonal lattices observed in the unstratified shear flow simulations beginning with three-dimensional random initial conditions by Comte et al. (1992), and the 'diamond' lattices noticed in the unstratified experiments and simulations of Nygaard and Glezer (1994) and Collis et al. (1994), using spanwise phase variations in the initial conditions. The vertical structures do not have much curvature, hence the vortex pairing proceeds to completion, in contrast to the observations of Collis et al. (1994).

Basak and Sarkar (2005) demonstrate that it is indeed buoyancy that brings about the lattice of vortex elements. The velocity fluctuations stir up the initial linearly-varying density field to create density fluctuations. In a vortex core, regions develop where the density field becomes segregated on a large scale resulting in a horizontal density gradient which tends to tilt these sections of the vortex core through a baroclinic torque. The tilting brings local regions of adjacent vortex cores closer, which are then able to pair locally. As a consequence, a network of quasi-vertical, truncated vortex cores connected by inclined segments emerges. These inclined segments collapse due to gravity and strain, with the former effect being increasing important when N is large. Finally, a lattice of organized vorticity emerges which is quite distinct from the hexagonal lattice observed in the unstratified low-Re studies of Comte et al. (1992) and Nygaard and Glezer (1994).

The three dimensional structure of the vertical and horizontal vorticity in case A5 is elucidated by plotting the respective vorticity isosurfaces at a late time, $t = 79.8$, between

$x = 38$ and 72. The organization of the vertical vorticity into a lattice of quasi-vertical, stubby vortices and of the horizontal vorticity into collapsed sheets is noteworthy. The thin sheets of horizontal vorticity occur in the vicinity of the dislocations of the columns of vertical vorticity. Some of the horizontal vorticity is associated with thin patches of intrusions of mixed fluid away from the shear layer. These thin regions of intense vorticity are likely to be the sites of strong dissipation of turbulent kinetic energy ϵ and mixing of the density field. Some of these patches also have associated vertical vorticity as can be seen in the lower part of Fig. 31(a). The vertical scale of these thin ω_h patches is expected to be related to the viscous scales, while their horizontal scale is related to the horizontal distance between the dislocations, which in turn scales roughly with the shear layer thickness δ. Case A2, at later time, also displays a similar three-dimensional structure. The main difference is that, in case A2, the isosurface has more small-scale wrinkling and that, in addition to the sheets of horizontal vorticity, there is significant ω_h in the braid regions between adjacent vortex cores as well as wrapped around the cores.

6 Concluding remarks

Stratification generally decreases turbulence levels in shear flows with the details being flow dependent. In uniform shear flow, stratification can cause turbulence to decay instead of an increase with time observed in the corresponding unstratified case. Such a collapse of turbulence has also been observed in the much-studied shear layer where the velocity profile is inflectional. However, in wall-bounded turbulence, the stratification and shear in the near-wall region are such that turbulence is maintained so that local turbulent transport and mixing are possible even if the overall stratification is very large. Horizontal mean shear is more effective than vertical mean shear of the same magnitude in mixing of the density field. This is primarily because horizontal eddying motions, not directly affected by stratification, are more energetic and are able to transfer energy to the vertical velocity component at small scales which accomplishes vertical mixing. The horizontal shear layer with an inflectional mean velocity profile is an interesting flow where buoyancy initially inhibits three-dimensional turbulence but later promotes vertical mixing by introducing dislocations in quasi two-dimensional vortex cores leading to layers of large fluctuating vertical shear. Although not discussed here, there are intrusions and waves at the edges of the shear layer that feed off the vortices and turbulence inside the shear layer, thus contributing to transport.

Identification of a measure of stratification that could be used to infer the properties of the flow have a long history. Linearized stability analysis of flows with vertical shear identifies the gradient Richardson number, Ri_g; if $Ri_g > 0.25$ globally, the flow is stable. Ri_g is clearly important in turbulent flows too. In uniform shear flow, Ri_g distinguishes between growing and decaying turbulence, in the shear layer it gives an estimate of the shear layer thickness at which turbulence collapses and, in a variety of flows with vertical shear, the maximum value of the flux Richardson number occurs at $Ri_g \simeq 0.25$. However, when stratification is imposed away from the region where turbulence is produced, for instance, the problem of open channel flow with adiabatic bottom wall and heating at the free surface discussed here, Ri_g is not the primary indicator of stratification effects and a bulk Richardson number or a Froude number may be more appropriate.

The ratio B/ϵ, the so-called efficiency of mixing (more correctly an efficiency of transport), was examined in all of these flows. In vertical shear flows, this ratio is a non-monotone function of Ri_g with the maximum value being approximately 0.2 and an eventual decrease with increasing stratification. In horizontal shear flows, the value of B/ϵ is generally larger than when the shear is vertical and, furthermore, there does not appear to be a large drop in its value at large Ri_g.

Bibliography

V. Armenio and S. Sarkar. An investigation of stably stratified turbulent channel flow using large-eddy simulation. *J. Fluid Mech.*, 459:1–42, 2002.

V. Armenio and S. Sarkar. Mixing in a stably-stratified medium by horizontal shear near vertical walls. *Theor. Comput. Fluid Dynamics*, 17:331–349, 2004.

S. P. S. Arya. Buoyancy effects in a horizontal flat-plate boundary layer. *J. Fluid Mech.*, 68:321, 1975.

S. Basak and S. Sarkar. Dynamics of a Stratified Shear Layer with Horizontal Shear. *submitted*, 2005.

J. H. Bell and R. D. Mehta. Development of a two-stream mixing layer from tripped and untripped boundary layers. *AIAA J.*, 28:2034–2042, 1990.

R. E. Britter. *An experiment on turbulence in a density stratified fluid.* PhD thesis, Monash Univerity, Australia, 1974.

S. Scott Collis, Sanjiva K. Lele, Robert D. Moser, and Michael M. Rogers. The evolution of a plane mixing layer with spanwise nonuniform forcing. *Phys. Fluids*, 6(1):381–396, 1994.

Pierre Comte, Marcel Lesieur, and Eric Lamballais. Large-scale and small-scale stirring of vorticity and a passive scalar in a 3-D temporal mixing layer. *Phys. Fluids*, 4(12): 2761–2778, 1992.

P. F. Crapper and P. F. Linden. The structure of turbulent density interfaces. *J. Fluid Mech.*, 65:45, 1974.

P. E. Dimotakis and G. L. Brown. Mixing layer at high reynolds-number - large-structure dynamics and entrainment. *J. Fluid Mech.*, 78:535, 1976.

D.M. Farmer, E.A. D'Asaro, M.V. Trevorrow, and G.T. Dairiki. Three-dimensional structure in a tidal convergence front. *Continental Shelf Research*, 15:1649–1673, 1995.

H. J. S. Fernando. Turbulent mixing in stratified fluids. *Ann. Rev. Fluid Mech.*, 23: 455–493, 1991.

P. Flament, R. Lumpkin, J. Tournadre, and L. Armi. Vortex pairing in an unstable anticyclonic shear flow: discrete subharmonics of one pendulum day. *J. Fluid Mech.*, 440:401–409, 2001.

K. S. Gage and W. H. Reid. The stability of thermally stratified plane Poiseuille flow. *J. Fluid Mech.*, 33:21, 1968.

R. P. Garg, J. H. Ferziger, S. G. Monismith, and J. R. Koseff. Stably stratified turbulent channel flows. I. Stratification regimes and turbulence suppression mechanism. *Phys. Fluids*, 12:2569, 2000.

T. Gerz, U. Schumann, and S. E. Elghobashi. Direct numerical simulation of stratified homogeneous turbulent shear flows. *J. Fluid Mech.*, 200:563–594, 1989.

T. Gerz H.-J. Kaltenbach and U. Schumann. Large-eddy simulation of homogeneous turbulence and diffusion in stably stratified shear flow. *J. Fluid Mech.*, 280:1–40, 1994.

M. A. Hernan and J. Jimenez. Computer-analysis of a high-speed film of the plane turbulent mixing layer. *J. Fluid Mech.*, 119:323, 1982.

S. E. Holt, J. R. Koseff, and J. H. Ferziger. A numerical study of the evolution and structure of homogeneous stably stratified sheared turbulence. *J. Fluid Mech.*, 237: 499–539, 1992.

E. J. Hopfinger. Turbulence in stratified fluids: A review. *J. Geophys. Res.*, 92:5287–5303, 1987.

L. N. Howard. Note on a paper of John. W. Miles. *J. Fluid Mech.*, 10:509–512, 1961.

G. N. Ivey and J. Imberger. On the nature of turbulence in a stratified fluid. Part I: the energetics of mixing. *J. Phys. Oceanogr.*, 21:650–658, 1991.

F. G. Jacobitz and S. Sarkar. The effect of nonvertical shear on turbulence in a stably stratified medium. *Phys. Fluids*, 10(5):1158–1168, 1998.

F. G. Jacobitz and S. Sarkar. On the Shear Number Effect in Stratified Shear Flow. *Theor. Comput. Fluid Dynamics*, 13:171–188, 1999a.

F. G. Jacobitz and S. Sarkar. A direct numerical study of transport and anisotropy in a stably stratified turbulent flow with uniform horizontal shear. *Flow, Turbulence and Combustion.*, 63:343–360, 1999b.

F. G. Jacobitz, S. Sarkar, and C. W. VanAtta. Direct numerical simulations of the turbulence evolution in a uniformly sheared and stably stratified flow. *J. Fluid Mech.*, 342:231–261, 1997.

J.A. Johannessen, R.A. Schuman, G. Digranes, D.R. Lyzenga, C. Wackerman, O.M. Johannessen, and P. W. Vachon. Coastal ocean fronts and eddies imaged with ERS 1 synthetic aperture radar. *J. Geophys. Res.*, 101:6651–6667, 1996.

S. Komori. *Turbulence structure in stratified flow.* PhD thesis, Kyoto Univerity, Japan, 1980.

S. Komori, H. Ueda, F. Ogino, and T. Mizushina. Turbulence structures in stably stratified open-channel flow. *J. Fluid Mech.*, 130:13–26, 1983.

B. Kosovic and J. A. Curry. A large eddy simulation study of a quasi-steady, stably stratified atmospheric boundary layer. *J. Atmos. Sci.*, 57:1052, 2000.

M. Lesieur. *Turbulence in fluids, 3rd edn.* Springer, 1997.

R. Lien and T.B. Sanford. Turbulence spectra and local similarity scaling in a strongly stratified oceanic bottom boundary layer. *Continental Shelf Research*, 24:375–392, 2004.

J. T. Lin and Y. H. Pao. Wakes in stratified fluids. *Ann. Rev. Fluid Mech.*, 11:317–338, 1979.

Y. Lu, R. G. Lueck, and D. Huang. The effect of stable thermal stratification on the stability of viscous parallel flows. *J. Phys. Oceanogr.*, 30:855–867, 2000.

L. Mahrt. Stratified atmospheric boundary layers. *Boundary-Layer Meteorology*, 90: 375–396, 1999.

P. J. Mason and S. H. Derbyshire. Large eddy simulation of the stably-stratified atmospheric boundary layer. *Boundary-Layer Met.*, 53:117, 1990.

J. W. Miles. On the stability of heterogeneous shear flows. *J. Fluid Mech.*, 10:496–508, 1961.

M. J. Moore and R. R. Long. An experimental investigation of turbulent stratified shearing flow. *J. Fluid Mech.*, 49:635–655, 1971.

P. Müller, G. Holloway, F. Henyey, and N. Pomphrey. Nonlinear interactions among gravity waves. *Rev. Geophys.*, 24:493–536, 1986.

W. Munk, L. Armi, K. Fischer, and F. Zachariasen. Spirals on the sea. *Proc. R. Soc. Lond. A*, 456:1217–1280, 2000.

R. Nagaosa and T. Saito. Turbulence structure and scalar transfer in stably stratified free-surface flows. *AIChE J.*, 43:2393, 1997.

K. J. Nygaard and A. Glezer. The effect of phase variations and cross-shear on vortical structures in a plane mixing layer. *J. Fluid Mech.*, 276:21–59, 1994.

Y. Pan and S. Banerjee. A numerical study of free-surface turbulence in channel flow. *Phys. Fluids*, 7:1649–1664, 1995.

W. R. Peltier and C. P. Caulfield. Mixing efficiency in stratified shear flows. *Ann. Rev. Fluid Mech.*, 35:135–167, 2003.

J. F. Piat and E. J. Hopfinger. A boundary layer topped by a density interface. *J. Fluid Mech.*, 113:411, 1981.

P. S. Piccirillo and C. W. VanAtta. The evolution of a uniformly sheared thermally stratified turbulent flow. *J. Fluid Mech.*, 334:61–86, 1997.

J. J. Riley and M. P. Lelong. Fluid motions in the presence of strong stable stratification. *Ann. Rev. Fluid Mech.*, 32:613–657, 2000.

R. S. Rogallo. Numerical experiments in homogeneous turbulence. *NASA TM 81315*, 1981.

M. M. Rogers, P. Moin, and W. C. Reynolds. *The structure and modeling of the hydrodynamic and passive scalar fields in homogeneous turbulent shear flow.* PhD thesis, Stanford University, Report TF-25, Mechanical Engr., 1989.

M. M. Rogers and R. D. Moser. Direct simulation of a self-similar turbulent mixing layer. *Phys. Fluids*, 6(2):903–923, 1994.

J. J. Rohr, E. C. Itsweire, K. N. Helland, and C. W. VanAtta. Growth and decay of turbulence in a stably stratified shear flow. *J. Fluid Mech.*, 195:77–111, 1988.

E. S. Saiki, C. H. Moeng, and P. P. Sullivan. Large-eddy simulation of the stably-stratified planetary boundary layer. *Boundary-Layer Met.*, 53:117, 1990.

U. Schumann and T. Gerz. Turbulent mixing in stably stratified shear flows. *J. Appl. Meteor.*, 34:33, 1995.

L. H. Shih, J. R. Koseff, J. H. Ferziger, and C. R. Rehmann. Scaling and parameterization of stratified homogeneous turbulent shear flow. *J. Fluid Mech.*, 412:1–20, 2000.

M. T. Stacey, S. G. Monismith, and J. R. Barua. Observations of turbulence in a partially stratified estuary. *J. Phys. Oceanogr.*, 29:1950–1970, 1999.

S. Tavoularis and U. Karnik. Further experiments on the evolution of turbulent stresses and scales in uniformly sheared turbulence. *J. Fluid Mech.*, 204:457–478, 1989.

J. Taylor, S. Sarkar, and V. Armenio. Large eddy simulation of stably stratified open channel flow. *submitted*, 2005.

K. B. Winters and E. A. DAsaro. Diascalar flux and the rate of fluid mixing. *J. Fluid Mech.*, 317:179–193, 1996.

K. B. Winters, P. N. Lombard, J. J. Riley., and E. A. DAsaro. Available potential energy and mixing in density-stratified fluids. *J. Fluid Mech.*, 289:115–128, 1995.

Chapter 4:
The atmospheric boundary layer

F.T.M. Nieuwstadt

J.M. Burgers Centre
Leeghwaterstraat 21, 2628 CA Delft, the Netherlands

Abstract A review is presented of the turbulent structure and dynamics of the atmospheric boundary layer. First attention is given to some general aspects of the boundary layer in relation to its role in the dynamics of the atmosphere. The general characteristics of the boundary layer, such as turbulence are discussed and the equations for a horizontally homogeneous boundary layer are formulated. Based on a simple closure of the turbulent fluxes a solution of these equations, in the form of the Ekman profile, is presented. The structure and dynamics of boundary layer prototypes, such as the surface layer, the neutral, the convective and the stable boundary layer, are investigated. Within each scaling regime the appropriate scaling parameters are defined and scaling expressions are formulated and compared with experimental data. The boundary-layer dynamics of each prototype is considered. In addition attention is given to some special topics such as the effect of roughness and vegetation, coherent structures and clouds. Finally the non-stationary and inhomogeneous boundary layer is discussed.

1 Introduction

The atmosphere can be considered as a heat engine fed by temperature difference between the equator and the poles. This temperature difference is the consequence of the radiation balance. Due to the position of the sun with respect to the earth, the solar elevation (e.g. at midday) is high in tropics but low at the poles with as a consequence that the solar energy[1] is deposited over a larger surface area at the poles that at the equator. The radiation of the sun is usually denoted as short wave radiation because the radiation intensity is roughly equal to the radiation from a black body at about 6000 °K, i.e. the temperature of the sun. On the other hand the earth radiates energy back into space by so-called long-wave radiation, which is related to the black body radiation at about the temperature of the earth surface. The radiation balance prescribe that the incoming solar radiation and the outgoing long wave radiation should be approximately in balance,

---

[1] The solar insolation is about 600 Watts m^{-2} when measured at the earth surface at an orientation perpendicular to the solar rays. The solar at the edge of our atmosphere is about 1350 Watts m^{-2}. The difference is due to reflection, also called albedo, by the surface, clouds and aerosol particles and by absorption, e.g. in the ozone layer.

if the earth is to be in a stationary state. However, the solar radiation strikes the surface at the equator at midday at about an angle of 90° while the angle at the poles is much smaller. The consequence that insolation per m² at the equator is larger than at the poles. On the other hand the outgoing long wave radiation is more or less independent of position at the globe. The consequence is that despite the overall radiation balance we have a local imbalance with more incoming than outgoing radiation at the equator and the reverse at the pole. This provides the fuel for the atmospheric heat engine, which takes form in the general circulation of the atmosphere with its weather systems and also for the large-scale motions in the oceans.

The strong solar insolation (about 600 Watts m⁻²) at the equator heats up the air directly at the surface by means of conduction and this type of heat transfer is in meteorology usually denoted as sensible heat heat flux. However, the solar insolation also results in evaporation, also known as latent heat flux because it results in heating the air a higher levels due to condensation leading to the strong showers that are so characteristic for the equatorial regions. The ratio of surface sensible to latent heat flux, which is known as the Bowen ratio, depends on the surface conditions in particular its moisture content. As a order magnitude the sensible heat flux above a standard surface is about equal to the latent heat flux. On the average the warm air at the equator rises and is then transported at these higher levels in the direction of the poles. At the same time at the surface this hot air is replenished by a flow of colder air from the direction of the poles. This circulation system is known as the Hadley cell. Around a latitude of about 30-40 ° a stable circulation can no longer be maintained due to the influence of the earth rotation[2] and the Hadley cell breaks up. As a consequence the air moving polewards at higher levels is forced down. Air that is moving down heats up and as a result the relative humidity goes down[3] or alternatively the air becomes drier. This is the reason that the 40-30 ° latitudes mark the location of the great deserts.

Beyond the Hadley cell, i.e. at the middle latitudes, the poleward transport of heat takes place in the form of large-scale weather systems, which take the shape of alternating high and low pressure systems. The process by which these pressure systems form and evolve is knows as baroclinic instability. It gives rise to strongly variable weather of the mid-latitudes. The atmospheric circular in the neighbourhood of the poles becomes weaker as here the need for polar transport of heat is no longer a driving force.

The circulation systems or large scale atmospheric motions, that we have mentioned above, take place in the so-called troposphere, which is a layer of air above the earth's surface with a thickness that measures about 11 km at the equator and about 8 km at the poles. The troposphere contains about 80% of the total mass of the atmosphere. Above the troposphere we find the stratosphere. The difference between the troposphere and stratosphere lies in their dynamics. While we have argued above that in the troposphere

---

[2]In tropical regions the earth's rotation in combination with the surface transport of air towards the equator is responsible for the trade winds.

[3]The relative humidity is the ratio between the specific humidity, which is the amount of water vapour in g per kg air, and the specific humidity where saturation takes place, which is a function of temperature. When air becomes warmer the saturation specific humidity goes up, while the amount of water vapour stays equal with as result that the relative humidity goes down.

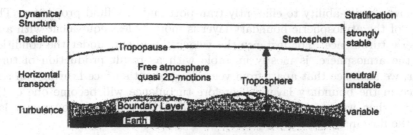

Figure 1. The general structure of the atmosphere. Reproduced with permission from Stull (1988).

the poleward transport of heat is responsible for the dynamics with as consequence horizontal but also vertical air motions, the dynamics of the stratosphere is primarily determined by radiation processes. This difference in dynamics results for troposphere in a decreasing temperature with height or alternatively in a neutral or slightly unstable atmosphere. In the stratosphere the temperature decreases much slower with height[4], which leads to stable conditions. As a result the flow in the stratosphere is laminar in contrast with the troposphere where the horizontal and vertical air motions cause turbulence. This fact has for instance the important consequence that the dispersion of pollutants in the stratosphere is much slower than in the troposphere. The height where the troposphere goes over into the stratosphere is known as the tropopause. In figure 1 we give an overview of the structure of the atmosphere that we have discussed above.

In this chapter we shall limit ourselves to the troposphere and within this troposphere we shall restrict our discussion even further to the layer of the atmosphere just above the earth surface. This layer is known as the atmospheric boundary layer. It is the region where the meteorological variables, such as wind velocity, temperature and humidity, adjust from their values in the free-atmosphere[5] to their values at the earth's surface. Alternatively it provides boundary conditions for the tropospheric flow in the form of the exchange of heat, moisture and momentum between the surface and the free atmosphere. The transport of heat and humidity is directly related to the solar insolation deposited at the earth surface resulting in the so-called sensible and latent heat flux. The transfer of momentum is related to the drag, which the air motions in the free atmosphere experience as a results of the presence of the earth surface because the boundary condition at surface requires that the relative airspeed with respect to the surface must be zero.

The theory of the atmospheric boundary layer should provide us with the vertical profiles of the meteorological variables, such as wind temperature and humidity, and their fluxes. It will be evident that these profiles must depend strongly on the process that determines the vertical transport and mixing in the boundary layer. This process is turbulence, which is known as a fluctuating and chaotic state of fluid motion and which

---

[4] At the top of the stratosphere, where the ozone layer starts, the temperature actually increases with height.

[5] From now on we shall denote the troposphere and the air motions in it by the free atmosphere.

is also known for its ability to efficiently transport and mix fluid properties. Therefore, the study of the atmospheric boundary layer is more or less equivalent with a study of atmospheric turbulence. For instance the velocity gradient, under the conditions that exist in the atmosphere, is mostly unstable with as result production of turbulence. Moreover, we shall see that apart from velocity gradient surface heating can also cause turbulence in the boundary layer. Therefore turbulence will become one of the main ingredients when we consider the structure and dynamics of the boundary layer. In contrast the flow in the free atmosphere is mostly laminar[6]

Another argument to justify a study of the atmospheric boundary layer is the fact that most of us live directly inside the boundary layer and we could alternatively call it the biosphere. For instance pollutants due to human activity disperses first in the atmospheric boundary layer before it reaches the troposphere or even the stratosphere. The fact that the flow within the boundary layer is turbulent promotes strong dispersion of these pollutants to concentration levels which can be tolerated without acute danger for our health. One might say that without this process of turbulence dispersion the biosphere would be unhabitable.

So let us define the atmospheric boundary layer as the thin layer near the earth's surface, in which the interaction between the surface and the free atmosphere, i.e. the troposphere above the boundary layer, takes place by means of the turbulent transport of momentum, heat and humidity. Our knowledge of atmospheric turbulence and of the atmospheric boundary layer has developed quite extensively over the last decades. In this review we can only present a limited selection of results that are now accepted as facts in the field of boundary-layer meteorology. However, several reviews and textbooks have appeared recently, to which an interested reader may turn for additional information. These are the reviews by Wyngaard (1988) and (1992) and the textbooks by Panofsky and Dutton (1984), Stull (1988), Arya (1988), Garratt (1992) and Kaimal and Finnigan (1994).

In the following sections we shall first derive the equations which govern the motion in the boundary layer. Thereafter, we shall introduce the various types of the atmospheric boundary layer and discuss their characteristics.

2 Basic equations

Air can be considered as a mixture of gases in combination with the three phases of water: vapour, liquid, and ice. The mixture of gases, of which the main components are nitrogen, oxygen, argon, and carbon dioxide, is what we call dry air . The equation of state for dry air follows the ideal gas law given by

$$\frac{p}{\rho} = R_d T \tag{2.1}$$

---

[6]The atmospheric boundary layer is defined here as being continuously turbulent. This distinguishes the boundary layer from the troposphere where the flow is in general only intermittently turbulent, e.g. in regions of large wind shear, which occur in convective storms that reach up into the troposphere, in breaking (mountain) waves or in the regions of high shear connected to the jet stream.

where p is the pressure, $\rho \approx 1.2$ kg m^{-3} the density, T the absolute temperature and R_d ($\simeq 287$ J kg^{-1} K^{-1}) the gas constant for dry air.

It is convenient to introduce the potential temperature (ϑ), which stays constant when a particle of dry air moves up or downwards isentropically. This means that the potential temperature is a conservative variable for vertical motion[7]. In the atmospheric boundary layer the vertical gradient of ϑ can be approximated by

$$\frac{d\vartheta}{dz} = \frac{dT}{dz} + \gamma_d \tag{2.2}$$

with $\gamma_d = g/c_p$ where g ($= 9.81$ m s^{-2}) is the acceleration of gravity, c_p ($= 1005$ J kg^{-1} K^{-1}) the specific heat of air at constant pressure and z the height above the earth's surface. The γ_d is also known as the dry adiabat. The gradient of the temperature dT/dz in relation to this dry adiabat determines the so-called static stability of the atmosphere. This means that when $dT/dz < -\gamma_d$ ($d\vartheta/dz < 0$) the atmosphere is absolute unstable, because a slight upward vertical displacement of an air parcel leads to a temperature of this parcel larger than its surrounding with the consequence so that it will rise further (an analogous behaviour is found for a downward displacement). The opposite case where $dT/dz > -\gamma_d$ ($d\vartheta/dz > 0$) the atmosphere is called stable because an vertically displaced particle will have a temperature lower (for an upward displacement) or higher (for a downward displacement) than its surrounding and thus will return to its initial position. The atmosphere is called neutral when $dT/dz = -\gamma_d$ ($d\vartheta/dz = 0$).

When moisture in the form of water vapour is present in the boundary layer, the temperature must be corrected to

$$T_v = T(1 + 0.61 q_v) \tag{2.3}$$

where T_v is called the virtual temperature and where q_v is the specific humidity defined as the mass of water vapour per unit mass of moist air. For this case of moist air the equation of state (2.1) remains valid when T_v is substituted for T. In other words, by introduction of T_v we take the presence of water vapour in air into account. We find that $T_v > T$ which is caused by the fact that the molecular weight of water is smaller than that of air. Similarly, we can introduce a virtual potential temperature ϑ_v which is defined by (2.2) with T_v substituted for T. In general the correction from T to T_v is small and therefore we shall in the following neglect this effect and use only the T and ϑ as absolute and potential temperature.

Next we consider the equations, which govern the flow and transport processes in the atmosphere. In the derivation it is usually assumed that the variations in density and temperature are small with respect to a constant reference state ρ_o and T_o, which leads to

$$-\frac{\rho}{\rho_o} \simeq \frac{T}{T_o} \simeq \frac{\vartheta}{T_o}. \tag{2.4}$$

This also implies that the variations in ρ and T are taken to be independent of pressure variations. This result follows from the fact that the velocity in the boundary layer is

---

[7]The the introduction of conservative variables will turn out to be convenient when we will parameterize vertical exchange processes.

much smaller than the velocity of sound and the vertical motions in the boundary layer
are much smaller that the scale height of the atmosphere, which is the height (≈ 5 km)
where the density has reduced to about half its value at the surface. These facts also
imply that the flow in the boundary layer can be considered as divergence free.

Moreover, we assume that the density variations only influence the flow dynamics
when combined with the acceleration of gravity, which is known as the Boussinesq approximation[8].

The equations for the conservation of mass, the conservation of momentum (also
known as the Navier-Stokes equation) and the conservation of energy expressed in terms
of the velocity vector u_i, the pressure p and the potential temperature ϑ in a Cartesian
coordinate system with the x_1 or x-axis pointing from West to East, the x_2 or y-axis
from South to North and the x_3 or z-axis oriented vertically with respect to the surface
then read

$$\frac{\partial u_i}{\partial x_i} = 0 \tag{2.5}$$

$$\frac{\partial u_i}{\partial t} + u_j \frac{\partial u_i}{\partial x_j} + 2\Omega \varepsilon_{ijk}\eta_j u_k = -\frac{1}{\rho_o}\frac{\partial p}{\partial x_i} + g\frac{\vartheta}{T_o}\delta_{i3} + \nu\frac{\partial^2 u_i}{\partial x_j^2} \tag{2.6}$$

$$\frac{\partial \vartheta}{\partial t} + u_j \frac{\partial \vartheta}{\partial x_j} = \kappa\frac{\partial^2 \vartheta}{\partial x_j^2}. \tag{2.7}$$

Here δ_{ij} is the Kronecker delta, ν ($= 1.5\ 10^{-5}$ m^2 s^{-1}) the kinematic viscosity and κ
($= 2.1\ 10^{-5}$ m^2 s^{-1}) the heat diffusivity. The terms $\partial u_i/\partial t + u_j\partial u_i/\partial x_j$ and $\partial\vartheta/\partial t +$
$u_j\partial\vartheta/\partial x_j$ denote the change of the velocity and temperature travelling along a trajectory
of a fluid particle. The term $2\Omega\varepsilon_{ijk}\eta_j u_k$ is the Coriolis acceleration due to the rotation of
the earth[9] with $\Omega = 2\pi/86400$ s^{-1} the angular velocity of the earth, ε_{ijk} the permutation
tensor and η_j the unit vector in the direction of the rotation axis of the earth. The
term $g\vartheta/T_o$ is know as the buoyancy term which describes the influence of temperature
variations (density variations) on the flow (note that this term plays only a role in the
equation for the vertical velocity component). The terms $\nu\partial^2 u_i/\partial x_j^2$ and $\kappa\partial^2\vartheta/\partial x_j^2$
describe the influence of molecular viscosity and diffusivity. A more extensive discussion
on the Boussinesq form of the governing equation can be found in chapter 2 (page 5).

2.1 Some aspects of turbulent flows

As already mentioned, turbulence[10] is generally known as a fluctuating and chaotic
state of flow motion. It is in principle is a solution of (2.5)-(2.7). However, such a
solution, which up to now has been only obtained by numerical computation, may not
be very practical because it gives us all the spatial and temporal details of the turbulent

---

[8]Velocity accelerations in the flow are in general small with respect to the acceleration of gravity
so that the effect of density variations in combination with these velocity accelerations can be
neglected in comparison with acceleration of gravity.

[9]The centrifugal acceleration is usually absorbed in the pressure term.

[10]We refer for more information on turbulent flows to Tennekes & Lumley (1972) and Pope
(2000)

flow whereas in general one is usually only interested in its average effect such as what are the turbulent transports of momentum, heat and humidity, let's say, averaged over one hour.

Therefore, it seems more appropriate to consider the statistics of a turbulent flow. To this end we introduce the following definition

$$u_i = \bar{u}_i + u_i' \quad \text{and} \quad \vartheta = \bar{\vartheta} + \vartheta',$$ (2.8)

by which the instantaneous velocity u_i and temperature ϑ is decomposed in a mean value, indicated by an overbar, and a fluctuation, indicated by a prime. The overbar denotes an ensemble average, i.e. an average over all turbulent flow states. In that case the average of the fluctuation becomes identical zero. Based on definition (2.8) we can define other statistics with as example

$$
\begin{aligned}
\overline{u_i^2} &= \overline{(\bar{u}_i + u_i')^2} \\
&= \bar{u}_i{}^2 + 2\overline{\bar{u}_i u_i'} + \overline{u_i'}{}^2 \\
&= \bar{u}_i{}^2 + \overline{u_i'}{}^2
\end{aligned}
$$ (2.9)

because $\overline{u_i'} \equiv 0$. Statistics such as $\bar{u}_i$ and $\overline{u_i'}{}^2$ are called one-point statistics because they are defined at one coordinate position and one time instant. The procedure introduce above is also known as Reynolds averaging and more details can be found in chapter 1 (page 6).

Applying decomposition (2.8) to the equations (2.5)-(2.7) leads to

$$\frac{\partial \bar{u}_i}{\partial x_i} = 0$$ (2.10)

$$\frac{\partial \bar{u}_i}{\partial t} + \bar{u}_j \frac{\partial \bar{u}_i}{\partial x_j} + 2\Omega \varepsilon_{ijk} \eta_j \bar{u}_k = -\frac{1}{\rho_o} \frac{\partial \bar{p}}{\partial x_i} + g \frac{\bar{\vartheta}}{T_o} \delta_{i3} + \nu \frac{\partial^2 \bar{u}_i}{\partial x_j^2} - \frac{\partial \overline{u_i' u_j'}}{\partial x_j}$$ (2.11)

$$\frac{\partial \bar{\vartheta}}{\partial t} + \bar{u}_j \frac{\partial \bar{\vartheta}}{\partial x_j} = \kappa \frac{\partial^2 \bar{\vartheta}}{\partial x_j^2} - \frac{\partial \overline{\vartheta' u_j'}}{\partial x_j}$$ (2.12)

where most terms have the same physical interpretation as we discussed in relation to equations (2.5)-(2.7). However, we see that some new terms appear, $\partial \overline{u_i' u_j'}/\partial x_j$ and $\partial \overline{\vartheta' u_j'}/\partial x_j$ and these are known as the Reynolds terms. They the describe the transport of momentum and heat by the turbulent motions. The $\overline{u_i' u_j'}$ and $\overline{\vartheta' u_j'}$ form two extra unknowns so that for a solution of the system of equations (2.10) - (2.12) two additional equations are needed to express these Reynolds terms as a function of the other unknowns $\bar{u}_i$, $\bar{p}$ and $\bar{\vartheta}$. This is known as the closure problem of turbulence.

The characteristics of the atmospheric boundary layer depend for a large part on the structure and dynamics of turbulence inside the boundary layer and in particular on which processes produce turbulence. This can be most conveniently studied by introducing the equation for the turbulent kinetic energy which reads

$$e = \frac{1}{2} \overline{\left(u'^2 + v'^2 + v'^2 \right)}$$ (2.13)

The equation for e can be derived from (2.5)-(2.7) with help of (2.10)-(2.12) and with as result

$$\underbrace{\frac{\partial e}{\partial t} + \bar{u}_j \frac{\partial e}{\partial x_j}}_{A} = \underbrace{-\overline{u'_i u'_j} \frac{\partial \bar{u}_i}{\partial x_j}}_{S} + \underbrace{\frac{g}{T_o} \overline{u'_i \vartheta'} \delta_{i3}}_{B} - \underbrace{\frac{\partial}{\partial x_j} \left(\overline{u'_j e'} + \frac{1}{\rho_o} \overline{u'_j p'} \right)}_{T} - \underbrace{\epsilon}_{D} \qquad (2.14)$$

with

$$e' = \frac{1}{2} \left(u'^2 + v'^2 + v'^2 \right).$$

The terms in this equation can be interpreted as: acceleration along a particle travelling with the mean velocity (A), shear production (S), buoyant production (B), turbulent and pressure transport, where we have neglected the viscous transport (T), and viscous dissipation (D) with following equation for the dissipation

$$\epsilon = \nu \overline{\left(\frac{\partial u'_i}{\partial x_j} \right)^2} \qquad (2.15)$$

In general the shear production term is always positive, which means that shear always produces turbulence. On the other hand the buoyant production can be positive and negative. When the buoyant-production term is positive, i.e. producing turbulence, we speak of a convective or unstable boundary layer and in the other case with the buoyant-production term negative the boundary layer is called stable. When the buoyant-production term is identically zero we call the boundary layer neutral. The characteristics of these different types of boundary layer will discussed further in section 4.

3 Boundary-layer equations

We have argued that the height of the boundary layer varies between 100 m to 1000 and this sets a limit to the vertical scale $\mathcal{L}_z$ of the flow in the boundary layer. On the other hand the scale of the horizontal motions $\mathcal{L}_x$ is usually much larger and lies in the order of 10 to 100 km so that $\mathcal{L}_z \ll \mathcal{L}_x$. The consequence of the difference in scales between horizontal versus vertical motions implies that $\partial /\partial x \equiv \partial /\partial y \ll \partial /\partial z$. If we substitute this scaling result in (2.5) we find that the vertical velocity $\bar{w} \ll \bar{u}, \bar{v}$ with

$$\frac{\bar{w}}{\bar{u}} \propto O \left(\frac{\mathcal{L}_z}{\mathcal{L}_x} \right).$$

Substitution of this scaling in (2.5) - (2.7) leads to a simplification, which is known as the boundary-layer approximation. With this approximation the resulting equations in their component form become

$$\frac{\partial \bar{u}}{\partial x} + \frac{\partial \bar{v}}{\partial y} + \frac{\partial \bar{w}}{\partial z} = 0 \qquad (3.1)$$

as the equation of continuity;

$$\frac{\partial \bar{u}}{\partial t} + \bar{u}_j \frac{\partial \bar{u}}{\partial x_j} - f\,\bar{v} = -\frac{1}{\rho_o}\frac{\partial \bar{p}}{\partial x} + \nu \frac{\partial^2 \bar{u}}{\partial z^2} - \frac{\partial \overline{u'w'}}{\partial z}$$

$$\frac{\partial \bar{v}}{\partial t} + \bar{u}_j \frac{\partial \bar{v}}{\partial x_j} + f\,\bar{u} = -\frac{1}{\rho_o}\frac{\partial \bar{p}}{\partial y} + \nu \frac{\partial^2 \bar{v}}{\partial z^2} - \frac{\partial \overline{v'w'}}{\partial z} \qquad (3.2)$$

$$2\Omega \cos \Phi \, \bar{v} = -\frac{1}{\rho_o}\frac{\partial \bar{p}}{\partial z} + g\frac{\bar{\vartheta}}{T_o} - \frac{\partial \overline{w'^2}}{\partial z}$$

as the conservation of momentum in the x, y and z-direction respectively[11] with f the Coriolis parameter defined as $f = 2\Omega \sin \Phi$ with Φ the latitude;

$$\frac{\partial \bar{\vartheta}}{\partial t} + \bar{u}_j \frac{\partial \bar{\vartheta}}{\partial x_j} = \kappa \frac{\partial^2 \bar{\vartheta}}{\partial z^2} - \frac{\partial \overline{\vartheta'w'}}{\partial z} \qquad (3.3)$$

as the conservation of energy expressed in terms of the potential temperature and

$$\frac{\partial \bar{q}}{\partial t} + \bar{u}_j \frac{\partial \bar{q}}{\partial x_j} = \kappa \frac{\partial^2 \bar{q}}{\partial z^2} - \frac{\partial \overline{q'w'}}{\partial z} \qquad (3.4)$$

as the conservation equation for the specific humidity q. In the following we shall consider the solutions of these equations for a rather small region (with respect to the size of the earth). This means that the latitude Φ and thus also f can be taken as approximately constant. This approximation is also known as the f-plane assumption.

The equations (3.1)-(3.4) form the point of departure in a description of the dynamics of the boundary layer and the transport of momentum, heat and humidity from the earth's surface to the free atmosphere.

3.1   General characteristics of the atmospheric boundary layer

The characteristics of the mean flow in the atmospheric boundary layer can be best examined by considering a simple case. For this we choose the stationary and horizontally homogenous boundary layer, i.e. we make the following, assumptions

$$\frac{\partial \,\overline{\cdots}}{\partial x} = 0, \quad \frac{\partial \,\overline{\cdots}}{\partial y} = 0 \quad \text{and} \quad \frac{\partial \,\overline{\cdots}}{\partial t} = 0.$$

Substituting these assumptions in the continuity equation (3.1) and using the boundary condition at the surface we find that $\bar{w} = 0$ for all x, y and z. The momentum equations (3.2) then reduce to

$$-f\,\bar{v} = -\frac{1}{\rho_o}\frac{\partial \bar{p}}{\partial x} - \frac{\partial \overline{u'w'}}{\partial z} \qquad (3.5)$$

$$f\,\bar{u} = -\frac{1}{\rho_o}\frac{\partial \bar{p}}{\partial y} - \frac{\partial \overline{v'w'}}{\partial z} \qquad (3.6)$$

---

[11]Note that the momentum equation in the z-direction reduces to a hydrostatic balance equations. This also allow a integration of these equations along the vertical coordinate resulting in what is known as the shallow water equations.

where we have omitted the equation for the vertical momentum, because it only results in a relation for the pressure as function of the height z. In the free atmosphere there is no turbulence so that the Reynolds stress becomes zero with as consequence that the Coriolis acceleration must be in equilibrium with the pressure gradient. This leads to the following result

$$-f\,V_g \;=\; -\frac{1}{\rho_o}\frac{\partial \bar{p}}{\partial x} \tag{3.7}$$

$$f\,U_g \;=\; -\frac{1}{\rho_o}\frac{\partial \bar{p}}{\partial y} \tag{3.8}$$

where the velocities U_g and V_g are known as the geostrophic wind components. If we neglect the dependence of the horizontal pressure gradients on height, the boundary layer equation (3.5) and (3.6) can be rewritten as

$$-f\,(\bar{v} - V_g) \;=\; -\frac{\partial \overline{u'w'}}{\partial z} \tag{3.9}$$

$$f\,(\bar{u} - U_g) \;=\; -\frac{\partial \overline{v'w'}}{\partial z} \tag{3.10}$$

We can proceed no further unless we provide a closure relationships for the Reynolds terms in (3.9) and (3.10). The closure problem in turbulence is a complicated subject and a fully satisfactory closure, i.e. a closure applicable to all conditions, is not available. For more information on turbulence models we refer to Durbin & Petterson (2000) and Launder & Sandham (2002). Here we shall use the most simple closure model, which assumes that the Reynolds terms can be treated in terms of a quasi-molecular model. The result reads

$$-\overline{u'w'} \;=\; K_m\frac{\partial \bar{u}}{\partial z} \tag{3.11}$$

$$-\overline{v'w'} \;=\; K_m\frac{\partial \bar{v}}{\partial z} \tag{3.12}$$

where K_m is called the eddy viscosity. The background of this closure relationship is the assumption that large-scale turbulent eddies are able to transport moment in a similar way as molecular motions transport momentum. Although there is no physical justification for this assumption, the closure relationships (3.11) and (3.12) perform in many cases quite satisfactory. Based on the assumption that the large-scale turbulent eddies take care of the momentum transport we can estimate the magnitude of the eddy viscosity K_m as

$$K_m \propto \ell\,v$$

where ℓ is the length scale of the large-scale eddies and v the velocity scale. Let us substitute this result in the ratio of K_m to the molecular viscosity ν. We then find

$$\frac{K_m}{\nu} = \frac{\ell\,v}{\nu} \equiv \mathrm{Re}_\ell$$

where Re_ℓ is the so-called turbulence Reynolds number. For a fully developed turbulent flow this Reynolds number is very large, which is especially the case for atmospheric turbulence. This implies that the eddy viscosity will be much larger than the molecular viscosity, which is consistent with the efficient mixing property of turbulence. In the following we shall denote the closure given by (3.11) and (3.12) as K-theory or gradient-transfer theory.

The next step is the specification of the eddy viscosity K_m or alternatively the length scale ℓ and the velocity scale v. A well-known approach, which goes back to Prandtl, is the so-called mixing-length hypothesis, which is given by

$$K_m = \ell^2 \left| \frac{\partial \bar{u}}{\partial z} \right| \tag{3.13}$$

where ℓ is called the mixing length with basically the same interpretation as discussed above, i.e. the length scale of the largest eddies.

With respect to the dependence of K_m on ℓ we note that near the surface the size of these eddies will be limited by the presence of the surface, i.e. the eddies can not penetrate through the surface. This implies that the size of the eddies becomes proportional to the height or $\ell \propto z$. Away from the surface the eddies can reach the maximum which as a first assumption can be taken proportional to the boundary-layer height.

Let us now assume for convenience that K_m is constant, i.e. we neglect the dependence on height z near the surface. The equations (3.9) and (3.10) can then be solved exactly with as boundary conditions that at the surface ($z = 0$) the velocity becomes equal to zero and that for $z \to \infty$ the velocity approaches the geostrophic wind, which we in this case assume to have only a component along the x-axis with magnitude G. The result then reads

$$\frac{\bar{u}}{G} = 1 - \cos\left(\sqrt{\frac{f}{2K_m}}z\right) e^{-\sqrt{\frac{f}{2K_m}}z} \tag{3.14}$$

$$\frac{\bar{v}}{G} = \sin\left(\sqrt{\frac{f}{2K_m}}z\right) e^{-\sqrt{\frac{f}{2K_m}}z}. \tag{3.15}$$

and this result is known as the Ekman profile after Norwegian scientist Ekman, who derived this solution in 1905 for the similar boundary layer, which develops in the upper layer of the oceans as a result of the stress exerted by the wind on the surface.

In figure 2 we show this solution in the form of a so-called hodograph, in which the two components of the velocity are displayed along the coordinate axes. We see that near the surface the velocity vector makes an angle of 45° to the left with respect to the geostrophic wind. In other words the wind direction turns to the right (i.e. veers) for increasing height.

The solution (3.14) also allows us to obtain expression for the boundary-layer height h. Let us define the boundary-layer height as the first level where the x-component of the velocity has the same direction as the geostrophic wind. The result becomes

$$h = \pi\sqrt{\frac{2K_m}{f}}. \tag{3.16}$$

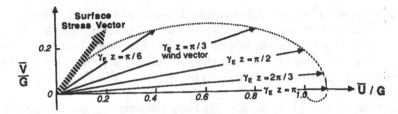

Figure 2. The theoretical Ekman profile .

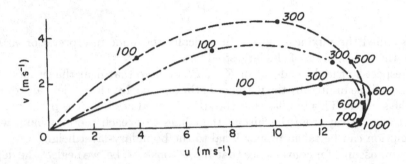

Figure 3. The observed velocity profile (solid line) in the atmospheric boundary layer and two theoretical Ekman profiles obtained with different assumptions for the eddy-viscosity profile.

With the order-of-magnitude estimates, $f \simeq 10^{-4}$ s^{-1} and $K_m \simeq 15$ m^2 s^{-1} we find that $h \simeq 1000$ m which is a realistic values considering measured values of the boundary-layer height. A perhaps more suitable definition of the boundary-layer height is the level where turbulence, which is responsible for the transport processes in the boundary layer, disappears. When the boundary layer develops non-obstructed, the height according to this latter definition is almost equivalent with the result given by (3.16). Both these definitions are not a very practical because in general we have neither the velocity nor the turbulence profile as a function of height. In many cases, however, the boundary-layer is obstructed from above by a so-called inversion, which is a region in which the (potential) temperature increases as a function of height[12]. We shall see later on that such a positive gradient of the (potential) temperature leads to a weakening of turbulence and even to its total disappearance when the inversion is strong enough. In that case the height of this inversion determines the vertical extent of the boundary layer and this height may be quite different from (3.16).

---

[12]Such an inversion may develop in a region of subsidence, which is an area where in the free atmosphere the air slowly descends, which is usually connected to a high pressure system. Convective turbulence in combination with a weak positive temperature gradient may also result in a strong inversion, which determines the top of the boundary layer. Finally, the cooling of the top of clouds by long wave radiation may also lead to a strong inversion.

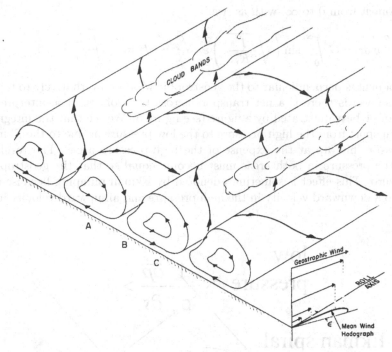

Figure 4. Roll vortices in the atmospheric boundary layer originating from an instability of the Ekman layer.

Although the Ekman profile (3.14) and (3.15) is an exact solution of the boundary-layer equations in the atmosphere, the fact is that it is almost never observed. This is even true for the Ekman profiles that are obtained with help of more realistic assumptions for the eddy viscosity. An example is for instance shown in figure 3 where the solid line illustrates the observed velocity profile while the dashed line gives the Ekman solution for a constant eddy viscosity. First we notice that the angle of the wind vector at the surface with the geostrophic wind is much smaller than the theoretical 45° and lies more in the neighbourhood of 10-20°. The dashed-dotted line, however, shows that the theoretical profile can be made to agree with the result for this angle by making the eddy viscosity K_m a function of height, for instance by taking into account that near the surface the size of the large-scale must become smaller (as discussed above). However, we also note that the shape of the observed profile is completely different from the theoretical solutions. The background of this difference lies in the fact that the theoretical Ekman profile is unstable due to the presence of an inflection point in the y-component of the velocity (Etling & Brown, 1999). This instability gives rise to large-scale roll vortices approximately aligned with the geostrophic wind as shown in figure 4.

Another property of the Ekman profile can be found by integrating the equation for the y-component from 0 to ∞, with as result

$$\int_0^\infty \bar{v}\, dz = G \int_0^\infty \sin\left(\sqrt{\frac{f}{2K_m}}z\right) e^{-\sqrt{\frac{f}{2K_m}}z} dz = \frac{G}{2}\sqrt{\frac{2K_m}{f}} > 0. \qquad (3.17)$$

The v-component is perpendicular to the geostrophic wind or alternatively to the isobars. Thus in other words there is a net transport across the isobars. The interpretation of this term can be best illustrated by a figure (see figure 5. We see that the integral (3.17) implies a transport from the high pressure to the low pressure or due to the Ekman layer the low-pressure fills up at the expense of the high-pressure area. The result is that eventually the pressure in both areas must become equal so that the geostrophic wind reduces to zero. This effect is sometimes denoted by Ekman pumping because it can be connected to a downward velocity in the high pressure and an upward velocity in the low pressure.

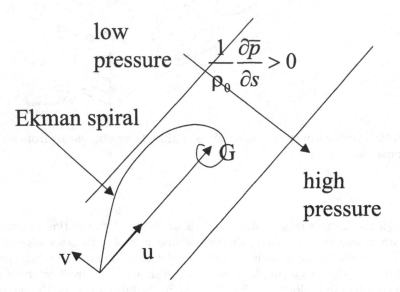

Figure 5. Illustration of the cross-isobaric transport by the Ekman profile.

Until now we have assumed (without explicitly mentioning this) that the geostrophic wind is independent of height. Let us now relax this assumption and consider the geostrophic wind as a function height. The result becomes

$$\frac{dU_g}{dz} = -\frac{1}{f\rho_o}\frac{\partial}{\partial y}\left(\frac{dp}{dz}\right) = \frac{g}{f\rho_o}\frac{\partial\rho}{\partial y} = -\frac{g}{fT_o}\frac{\partial T}{\partial y} + \frac{g}{fp_o}\frac{\partial p}{\partial y} \qquad (3.18)$$

$$\frac{dV_g}{dz} = \frac{1}{f\rho_o}\frac{\partial}{\partial x}\left(\frac{dp}{dz}\right) = -\frac{g}{f\rho_o}\frac{\partial\rho}{\partial x} = \frac{g}{fT_o}\frac{\partial T}{\partial x} - \frac{g}{fp_o}\frac{\partial p}{\partial x} \qquad (3.19)$$

where we have used the (simplified) hydrostatic relation: $\partial\bar{p}/\partial z = -\rho g$ and the ideal gas law (2.1). An order-of-magnitude estimate of the last term in both (3.18) and (3.19)

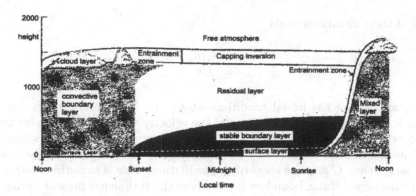

Figure 6. The daily cycle of the atmospheric boundary layer. reproduced with permission from Stull (1988).

leads to the result that this term can be neglected so that we find

$$\frac{dU_g}{dz} = -\frac{g}{fT_o}\frac{\partial T}{\partial y} \tag{3.20}$$

$$\frac{dV_g}{dz} = \frac{g}{fT_o}\frac{\partial T}{\partial x} \tag{3.21}$$

which are called the thermal wind relationships. So we see that a horizontal temperature gradient implies that the (geostrophic) wind changes with height where the direction of the temperature gradient determines in what direction the wind changes. In the introduction we have argued that the the heat engine of the atmospheric is fuelled by the temperature difference between the equator and the poles. Let us estimate the resulting temperature gradient as $\partial T/\partial y \simeq \Delta T_{equator-pole}/(0.25\pi R)$, which with $\Delta T_{pole-equator} \simeq -60°$ and with R the radius of the earth $\simeq 6000$ km, which leads to $\partial T/\partial y \simeq -0.01°$ km^{-1}. This result when substituted in (3.20), leads with $f \simeq 10^{-4}$ s^{-1} and $g \simeq 10$ m s^{-2} to $dU_g/dz \simeq 3$ m s^{-1} km^{-1}. This implies that at about 10 km height the wind speed would be about 100 km hour^{-1} to the east. This high-speed wind, which is indeed found at this height in the mid latitudes, is known as the jet stream and the instability phenomena, which go together with this jet, determine for a large part the weather in the mid latitudes.

The discussion above has been restricted to a stationary boundary layer. Let us now introduce time variation. First we introduce time dependence in the equations (3.5) and (3.6) when at the same time turbulence is neglected. This leads to

$$\frac{\partial v}{\partial t} - f\,\bar{v} = 0$$

$$\frac{\partial u}{\partial t} + f\,(\bar{u} - G) = 0$$

where we have assumed again that the geostrophic wind is aligned with the x-axis. The solution of these equations reads

$$\begin{aligned}
\bar{u} &= G + (\bar{u}_o - G)\, e^{ift} \\
\bar{v} &= \bar{v}_o\, e^{ift}
\end{aligned}$$

(3.22)

where $\bar{u}_o$ and $\bar{v}_o$ are the initial conditions at $t = 0$. This result, when depicted in a hodograph, represents a circular motion of the velocity vector with the geostrophic wind at the centre and with an period of $2\pi/f \simeq 17$ hours. This result is sometimes also denoted as the inertial circulation. The consequence of (3.22) is that the velocity can become larger than G at some stage of the oscillation. This is sometimes happens at the top of the nocturnal stable boundary layer, where the turbulence present during day time has decayed. As a result the wind at the top of this boundary layer exhibits a maximum which is also denoted as a low-level jet.

Another example of time variation in the atmospheric boundary layer is the daily cycle, which is the change of characteristics of the boundary layer over a land surface as a function of the time of day. As we shall see in section 4, during the day time the boundary layer is thick with strong turbulence while during night a stable shallow boundary layer develops with weak turbulence. This daily cycle is illustrated in figure 6 which will be used as guide for our discussion of the following sections.

3.2 Scaling

As we have argued above, the characteristics of the atmospheric boundary layer are determined for a large part by the structure and dynamics of the turbulence in the boundary layer. One of the most simple and at the same time successful tools to study problems in turbulence is dimension analysis, which is based on the introduction of dimensionless parameters. These parameters are usually defined in terms of appropriate length, velocity and time scales that are based on the various production processes of turbulence, which as we have seen in section 2.1 are shear or buoyancy. With the help of these dimensionless parameters we define so-called scaling regimes as those regions of the boundary layer, which are dominated by the influence of a certain turbulent process. This approach allows us to introduce some simplifications or approximations in our description of atmospheric turbulence.

Let us again restrict ourselves to the stationary and horizontally homogeneous boundary layer, which as we have discussed above is described by the equations (3.9) and (3.10). In figure 7 we show a schematic picture of this boundary layer together with all parameters that are possibly of importance for its structure and dynamics. These parameters are the geostrophic wind G (including of course its direction), the thermal wind dG/dz (and also its direction), the temperature jump $\Delta\theta$ which frequently develops at the top of a (unstable) boundary layer, the temperature gradient γ in the free atmosphere above the boundary layer, the Coriolis parameter f, the boundary-layer height h, the buoyancy parameter g/T_o ($\simeq 3.5\,10^{-2}$ m s^{-2} K^{-1}), which can be interpreted as the vertical acceleration experienced by a temperature fluctuation of 1 K, the roughness length z_o, which will discussed later in section 4.1 and finally the parameters τ_o and H_o, which are connected to the processes that produce turbulence in the boundary layer.

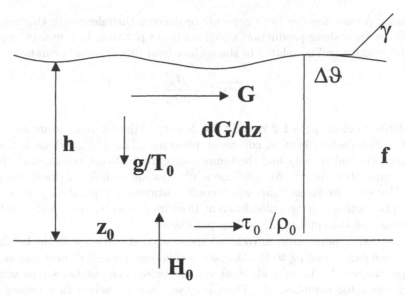

Figure 7. The atmospheric boundary layer and its characteristic parameters

To interpret τ_o and H_o we return to (2.14) which shows that there are two production processes. The first is shear production. It is due to the fact that the velocity near the surface must vary as a function of height to obey the no-slip boundary condition. This velocity gradient is almost always unstable under the conditions that exist in the atmosphere, and consequently turbulence is produced. Another way of looking at the no-slip condition is that it imposes a shear stress at the surface which is denoted by τ_o. It seems appropriate to scale the turbulence caused by this shear stress in terms of τ_o. Alternatively, we can also introduce a velocity scale related to τ_o, which is defined as

$$u_* = \sqrt{\frac{\tau_o}{\rho_o}} \tag{3.23}$$

and which is known as the friction velocity.

The second production process is buoyancy. It is related to density or temperature fluctuations, which, in the case of non-saturated air, are primarily due to the sensible heat flux imposed at the surface, where the background of this heat flux has been discussed in the introduction. From (2.14) it follows that a positive temperature flux, which occurs when heat is introduced at the surface will generated turbulence. The physical explanation is that the air heated at the surface is less dense than its surrounding and therefore tends to rise in the earth's gravitational field. So kinetic energy is produced. This process is known as free convection. However, the surface temperature flux can be also negative, i.e. heat is removed. The air near the surface will then be cooled and consequently its density increases. This situation is characterized as stable because the dense air will resist vertical motion and as a result turbulent motion is suppressed. In that case, the buoyant production term in (2.14) becomes negative. We thus see that the buoyancy production can both generate or destroy turbulence. In this respect, it is quite different from shear production which is always positive. In non-saturated air, the buoyant production will be related to the surface heat flux H_o which can be rewritten as

$$\overline{w'\vartheta'}_o = \frac{H_o}{\rho_o c_p} \tag{3.24}$$

with, as defined before, $\rho_o \simeq 1.2$ kg m^{-3} the density of the reference state and $c_p \simeq 1000$ J kg^{-1} K^{-1} the specific heat at constance pressure. The $\overline{w'\vartheta'}_o$, which is the correlation between vertical velocity and the temperature fluctuations, is named as the surface (virtual) temperature flux[13]. We shall use $\overline{w'\vartheta'}_o$ to scale turbulence due to buoyant production. However, we stress that, when condensation or evaporation occurs, the local buoyancy production may be quite different from its surface value and the surface value may no longer be the appropriate scaling parameter.

With the scaling parameters introduced above in combination with the height z above the surface we can according to Buckingham's theorem define 8 dimensionless combinations of parameters[14]. To take all these dimensionless combinations into account will be clearly much too complicated. Therefore, we limit ourselves to a subset and thus neglect the influence of the other parameters. Before introducing the remaining scaling parameters, we first return to the parameters that have been connected above to the two turbulence production processes, i.e. u_* and $\overline{w'\vartheta'}_o$. Based on these parameters we can define a length scale L given by

$$L = -\frac{u_*^3}{k\frac{g}{T_o}\overline{w'\vartheta'}_o} \tag{3.25}$$

where k is the von-Karman constant, which will be discussed further in section 4.1. The L is known as the Obukhov length scale. It can be interpreted as the height above the surface where buoyancy production starts to dominate over shear production. This interpretation is the background for the dimensionless combination z/L, of which the relevance will become clear in section 4.1.

The introduction of the length scale L as a parameter to characterize the relative importance of the two production processes of atmospheric turbulence may suggest a search for other appropriate length scales in order to define dimensionless combinations. It is known that the macroscopic structure of turbulence scales with the geometry in which the turbulence occurs. In our case, this geometry is given by the boundary-layer height

$$h$$

which we have defined as the depth of the turbulent layer above the surface.

---

[13]The index v to denote virtual temperature has been omitted because in non-saturated air the contributions of the temperature fluctuations dominate over that of the humidity fluctuations

[14]We should note that we have omitted the viscosity ν and the heat diffusivity κ as parameters because the Reynolds number of the atmospheric boundary-layer flow is so large that molecular effect may be neglected

Convective boundary layer $(\overline{w\theta}_o > 0)$ Stable Boundary Layer $(\overline{w\theta}_o < 0)$

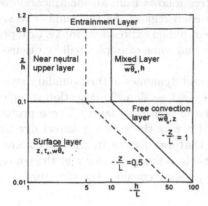 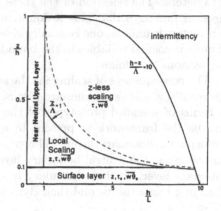

Figure 8. Scaling regions in the unstable $(\overline{w'\vartheta'}_o > 0)$ (left) and stable $(\overline{w'\vartheta'}_o < 0)$ (right) boundary layer; in each scaling region the characteristic parameters have been indicated (the buoyancy parameter (g/T_o) has been omitted as scaling parameter); the lines denote the boundaries between the various scaling regimes. Reproduced with permission from Holtslag and Nieuwstadt (1986)

Finally, another length scale that must play a role, is the height

$$z$$

above the surface.

With the three length scales introduced above, we can define two independent dimensionless combinations[15]. For this we may take

$$\frac{z}{h} \quad \text{and} \quad \frac{h}{L}.$$

The parameter z/h gives the location in the boundary layer and the parameter h/L can be interpreted as a stability parameter because, as we have seen above, $h/L < 1$ means that shear production dominates in the whole boundary layer while $h/L > 1$ implies that the (upper part of the) boundary layer is governed by the effect of buoyancy. A value of $h/L < 0$ implies production of turbulence and for the case that this production process dominates (i.e. $-h/L \gg 1$), we call the boundary layer convective. When $h/L > 0$, we have buoyant destruction of turbulence, and we call the boundary layer stable. For a value of h/L around zero, density effects will have small influence on the structure and dynamics of the boundary layer. This case is identified as a neutral boundary layer.

We can use both parameters h/L and z/h to define scaling regimes in the boundary layer. These scaling regimes are illustrated in figure 8. They can be interpreted as regions

---

[15]For the moment we shall neglect the roughness length z_o as a length scale to be taken into account.

where boundary-layer turbulence can be characterized in terms of a limited number of parameters. The selection of the these parameters follows from an identification of the physical processes that play a dominant role in each region. As the number of scaling parameters is limited, one is usually able to find simplified expressions for vertical profiles of meteorological variables in the boundary layer and some examples will be discussed in the sections that follow.

The consequences of scaling for the structure and dynamics of the boundary layer and the resulting scaling regimes are discussed in the next section. This section is organized in terms of so-called prototypes of the atmospheric boundary layer. These prototypes give us the framework to present in a systematic way the existing knowledge on the atmospheric boundary layer. The prototypes, that we discuss in this paper are: the surface layer, the neutral boundary layer, the stable boundary layer and the convective boundary layer. Of these boundary-layer types, we shall consider the structure in terms of scaling relationships and their dynamics.

4 Prototypes of the atmospheric boundary layer

Let us limit ourselves again to the horizontally homogeneous boundary layer, for which the horizontal derivatives on the left-hand side of the boundary-layer equation (3.2) can be neglected. The profiles in the horizontally homogeneous boundary layer are thus only a function of height. First, we consider for each prototype of the boundary layer the characteristic scaling parameters and illustrate how we can use these to formulate the structure in terms of profiles of various boundary-layer variables. Second, we consider the dynamics of each boundary-layer.

To discuss of boundary-layer dynamics, we need an estimate of the various time scales that can influence the time-dependent processes in the boundary layer.

Let us first consider the so-called forcing time scale

$$\mathcal{T}_f$$

which gives the time scale of the external processes that act upon the boundary layer and influence its development. An example is for instance the Coriolis parameter f, which would result for $\mathcal{T}_f$ in the inverse of the Coriolis parameter, which at mid-latitudes is equal to about $f \simeq 10^{-4}$ s^{-1}. However, the forcing time scale $\mathcal{T}_f$ also can be interpreted as the time variation of the surface boundary conditions such as the surface heat and humidity flux, which leads to a time scale of about 6 hours. Another possibility is the synoptic time variation of the large-scale pressure gradient or geostrophic wind, which forces the flow in the boundary layer.

The second time scale is

$$\mathcal{T}_m$$

which is related to the development of the mean boundary-layer structure, e.g. the mean velocity, humidity, or temperature profile. The boundary layer is called to be in equilibrium when $\mathcal{T}_m \propto \mathcal{T}_f$. In that case the parameters, which determine the structure and dynamics of the boundary layer, can be directly coupled to the external forcing parameters. When $\mathcal{T}_m \ll \mathcal{T}_f$ the boundary layer is called quasi stationary, which means that the time derivatives can be dropped from the boundary layer equations (3.2). The time evolution is then completely described by the time variation of the forcing parameters, which determine the structure of the boundary layer. Finally, when $T_m \gg T_f$ the dynamics of the boundary layer is primarily determined by internal parameters, which are for instance connected to the turbulence. In this latter category falls the nonstationary and inhomogeneous boundary layer, which we will discuss in section 5.3.

Next we introduce the time scale of the large-scale eddies or the macroscopic structure of turbulence in the boundary layer, which is given by

$$T_t \simeq \frac{\ell}{v}$$

where ℓ and v are a representative length and velocity scale for this macroscopic structure. These large-scale eddies are representative for the eddies responsible for the turbulent transport of momentum, heat and humidity.

Finally, we introduce

$$T_p$$

which denotes the time scale, at which the production process of turbulence, i.e. shear or buoyancy production, vary. For the case that $T_p \gtrsim T_t$ we call the turbulence quasi-stationary. In that case the time derivation of the turbulent kinetic energy on the left-hand side of (2.14) can be neglected. As the transport terms in this equation are usually of minor importance, it means that the production terms are in balance with the dissipation term or memory effects can be neglected. Such a balance: "production = dissipation" is in many cases a very good first-order assumption to describe the dynamics of turbulence. It leads for instance directly to the K-theory, which we have discussed as a closure assumption in section 3.1.

The complementary condition

$$T_t > T_p$$

is called strong non-stationarity. The development of turbulence is then governed by internal processes and memory effects become important. This evolution continues until the turbulence is adjusted to the new production processes and the condition "production = dissipation" becomes valid again. An example is for instance the condition around sunset when buoyant production quickly disappears and turbulence has to adjust to the new condition based. Other examples are the situation when turbulent flow is advected quickly over a variation in production, such as e.g. a land-sea transition.

4.1 Surface layer

Scaling We have defined the surface layer as the lower part of the boundary layer close to the surface, i.e. $z/h \leq 0.1$ (see figure 8). Given this definition we assume that the length scale h does not play a role for the structure of the surface layer. The background for this assumption is that near the surface the turbulent length scale will be determined by the distance to the surface rather than by the boundary-layer height. Moreover, we limit ourselves to

$$\frac{z}{z_0} \gg 1$$

where the roughness length z_0 characterizes the height of the roughness elements on the surface. This condition implies that z_0 will not influence directly the mean structure of the surface layer.

Given the scaling parameters introduced in section 3.2 we find that only

$$u_* \quad L \quad \text{and} \quad z$$

remain for scaling the structure of the surface layer. Our hypothesis is now that all turbulent one-point statistics[16] scaled in terms of these parameters follow a unique relationship as a function of z/L. This is called Monin-Obukhov similarity where the ratio z/L can be interpreted as a stability parameter. When $z/L \approx 0$, stability effects can be neglected. We see that this occurs for small values of z or for small values of $\overline{w'\vartheta'}_o$ with respect to u_*. In other words, buoyancy effects become always unimportant when we are sufficiently close to the ground and sometime everywhere but then $\overline{w'\vartheta'}_o$ needs to be small.

Let us consider as an example of Monin-Obukhov similarity the so-called flux-gradient relationships. These are the dimensionless velocity and temperature gradients defined by

$$\phi_m\left(\frac{z}{L}\right) = \frac{kz}{u_*}\frac{d\bar{u}}{dz}$$

$$\phi_h\left(\frac{z}{L}\right) = \frac{kz}{\vartheta_*}\frac{d\bar{\vartheta}}{dz} \qquad (4.1)$$

where $\bar{u}$ and $\bar{\vartheta}$ are the mean velocity and (potential) temperature in the surface layer and where the temperature scale ϑ_* is defined as $-\overline{w'\vartheta'}_o/u_*$ and k the von-Karman constant. In figure 9 we show a large number of observations plotted according to Monin-Obukhov similarity. The fact that the observations lie on a single curve can be taken as indirect proof for the validity of this scaling approach.

Based on data shown in figure 9, we can already learn some facts about the influence of stability on turbulence. For $z/L < 0$, which implies unstable or convective conditions, we have argued that turbulence is generated by buoyancy. We may expect in that case a large turbulence intensity and thus strong vertical transport. This will result in small gradients of the mean profiles and consequently in small values of the gradient functions ϕ_m and ϕ_h. The opposite is the case for stable condition when $z/L > 0$. Turbulence is then reduced by buoyancy effects and vertical transport becomes small. As a result, the mean gradients and consequently the values of ϕ_m and ϕ_h will be large.

Given the closure relationship (3.12), the expressions for ϕ_m and ϕ_h can be alternatively interpreted as the definition of an eddy viscosity and diffusivity. With $u_*^2 \equiv \tau_o/\rho_o$ and $\overline{w'\vartheta'}_o$ as the turbulent fluxes in the surface layer, which can assumed to be approximately constant, the expressions (4.1) can now be rewritten as

$$-\overline{u'w'} = K_m\frac{\partial\bar{u}}{\partial z}$$

$$-\overline{w'\vartheta'} = K_h\frac{\partial\bar{\vartheta}}{\partial z}$$

---

[16]Defined in connection with (2.9). See also chapter 1 of the present volume for a further discussion on one-point statistics.

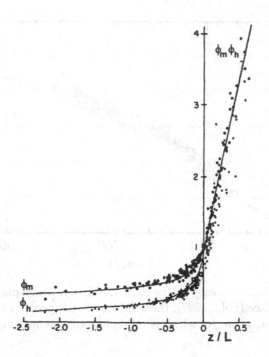

Figure 9. The flux relationships ϕ_m and ϕ_h in the surface layer scaled according to Monin-Obukhov similarity where ζ stands for z/L; the lines denote empirical expressions such as discussed in the text and the symbols are observations. Reproduced with permission from Fleagle and Businger (1980).

with

$$K_m = \frac{ku_*z}{\phi_m(z/L)} \quad \text{and} \quad K_h = \frac{ku_*z}{\phi_h(z/L)}. \tag{4.2}$$

In other words, the turbulent fluxes in the surface layer can be parameterized in terms of so-called K-theory. We shall return to the consequences of this result in the next subsection on the dynamics of the surface layer.

The functions ϕ_m and ϕ_h are still unknown. They must follow from fitting expressions to data such as those given in figure 9. Several empirical expressions have been proposed in the literature (see e.g. Sorbjan (1989) p. 74-76). An example is

$$\frac{z}{L} < 0 \qquad \phi_m = (1 - 16\frac{z}{L})^{-1/4} \qquad \phi_h = (1 - 14\frac{z}{L})^{-1/2}$$

$$\frac{z}{L} > 0 \qquad \phi_m = 1 + 5.2\frac{z}{L} \qquad \phi_h = 1 + 5.2\frac{z}{L}. \tag{4.3}$$

Let us now consider in somewhat more detail the result for ϕ_m at $z/L = 0$, i.e. at

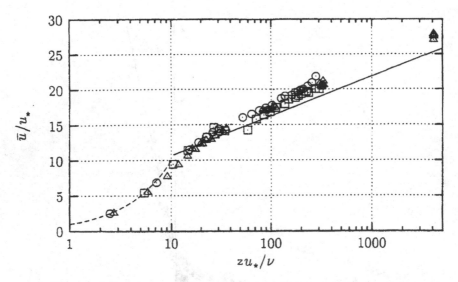

Figure 10. The logarithmic velocity profile measured in the atmospheric boundary layer over a very smooth surface (Folz, 1997); the solid line is the profile (4.5) and the dashed the linear the profile $\bar{u}/u_* = zu_*/\nu$, which is valid close to a smooth surface. Reproduced with permission.

neutral conditions. Both figure 9 and the empirical relations (4.3) show that[17] $\phi_m(0) = 1$. The flux-profile relationship can then be easily integrated with as result

$$\bar{u} = \frac{u_*}{k} \ln\left(\frac{z}{z_0}\right) \tag{4.4}$$

where we have used the boundary condition $\bar{u} = 0$ for $z = z_0$. This is the well-known logarithmic velocity profile which is assumed to be a fundamental property of every near-wall turbulent flow.

As we have already mentioned above the value of the roughness length z_0 depends on the characteristics of the surface. For a very smooth surface, e.g. a water surface under calm conditions or a flat ice surface, the z_0 becomes equal to

$$z_0 = \frac{\nu}{u_*} e^{-2}$$

so that the velocity profile can be written as

$$\bar{u} = \frac{u_*}{k}\left[\ln\left(\frac{z\,u_*}{\nu}\right) + 2\right]. \tag{4.5}$$

This equation has been verified in many laboratory experiments on engineering boundary layers but also for pipe and channel flows. It has even been verified in some atmospheric experiments carried out on very smooth salt flats as illustrated in figure 10.

---

[17]The fact that $\phi_m(0) = 1$ can be interpreted as a definition of the von-Karman constant k.

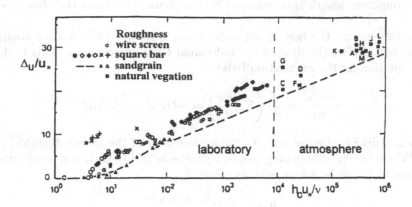

Figure 11. The function Δ_u as function of the roughness Reynolds number $u_* h_c/\nu$. Reproduced with permission from Raupach *et al.*, (1991)

However, in most conditions, which especially pertain to the atmosphere, the surface should be considered as rough. In this case the logarithmic profile (4.4) applies. In engineering applications the profile (4.5) for the smooth surface is modified to a rough surface following the equation

$$\frac{\bar{u}}{u_*} = \frac{1}{k} \ln \left(\frac{u_* (z-d)}{\nu} \right) - \left\{ \frac{1}{k} \ln \left(\frac{u_* z_0}{\nu} \right) - \frac{\Delta_u \left(\frac{h_c u_*}{\nu}, \lambda \right)}{u_*} \right\} \tag{4.6}$$

where the roughness function Δ_u depends on the height of the roughness elements h_c and where λ is called a roughness index, which describes the spatial distribution of the roughness elements. An example of the function Δ_u as a function of the Reynolds number $u_* h_c/\nu$ is shown in figure 11. When $u_* h_c/\nu \to 0$ the profile for the smooth wall should be retrieved whereas for $u_* h_c/\nu \to \infty$ relation (4.6) should reduce to (4.4), which is representative for a fully rough wall.

The variable d in (4.6) is known as the displacement length and it can be interpreted as a virtual origin in the logarithmic profile. This applies for instance to the case when the surface elements are so densely packed, e.g. for a grass covered surface, that they, so to speak, screen surface from the flow above the surface elements. The logarithmic profile is then lifted upward over a height equal to d. An extreme example is a tree canopy (see figure 12) or a built-up urban area, which consist of structures all of about the same height. Above the canopy the velocity adjusts to the logarithmic profile[18], in

---

[18] The turbulence structure above the roughness elements depends also on the spatial distributions of the roughness elements. When these are for instance widely placed, i.e. with some intermediate space, the turbulence structure and thus also the mean velocity profile is more representative for a mixing layer than for a wall-bound boundary layer.

which the roughness length is determined by the characteristics of the obstacles or tree tops.

Within the canopy the flow is not only determined by the boundary condition at the surface but also by the drag of the individual trees or obstacles. This leads to the following equation for the momentum balance

$$-\frac{d\overline{u'w'}}{dz} \equiv \frac{d}{dz}\left(K_m\frac{d\bar{u}}{dz}\right) = -D_f \equiv -\frac{A_f}{2}C_D\bar{u}^2 \qquad (4.7)$$

where C_D is a drag coefficient and A_f the surface area of the trees (obstacles) per unit volume. When for K_m the mixing length hypothesis (3.13) is chosen as closure model the equation (4.7) can be solved with as solution

$$\bar{u} = \bar{u}\left(h_c\right)e^{-\alpha(1-z/h_c)}$$

where the constant α is approximately equal to $\left[\left(A_f C_D h_c^3\right)/\left(2\ell_c^2\right)\right]^{1/3}$ and where ℓ_c is the mixing length in the canopy. The resulting canopy profile together with the logarithmic velocity profile above the canopy results in a profile with an inflection point, i.e. a point where the curvature of the profile changes from convex to concave. Such an inflection point is very unstable and leads to a break-up of the velocity into large-scale vortices, which dominate the flow and the vertical fluxes inside the canopy layer.

Another result, which is usually discussed, in connection with the logarithmic velocity profile, is the value of the von-Karman constant. In the derivation of the logarithmic velocity profile the von-Karman constant is usually assumed to be a universal constant, for which the value was usually taken to be $k = 0.4$. However, since the time of the Kansas experiments (Wyngaard, 1992), which were carried out around 1971 to verify experimentally the Monin-Obukhov similarity for the surface-layer and which resulted in a value of $k = 0.35$, the von-Karman constant and its value has been an active topic of investigation. In figure 13 we for instance illustrate the values of k which have been found by several experimenters over the last decades. The most recent result even claims the von-Karman constant to be a function of the roughness Reynolds number. This uncertainty about the value of the von-Karman constant is directly connected to the

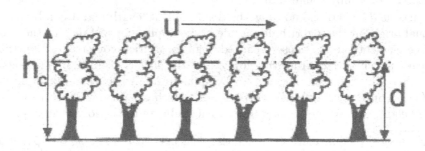

Figure 12. Schematic view of a forest canopy.

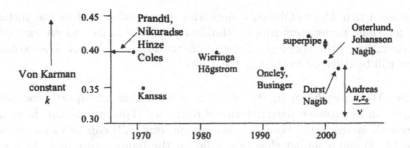

Figure 13. An overview of the various values of the von-Karman constant as found by various authors over the last decades.

recent questions that have been raised about the validity of the logarithmic velocity profile.

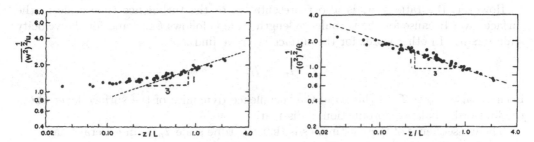

Figure 14. The standard deviation of the vertical velocity fluctuations (left) and of the temperature fluctuations (right) as function of z/L according to Monin-Obukhov similarity; the dashed line denote the free convection limit to be discussed in the section 4.3 on the convective boundary layer; the symbols denote experiments (Wyngaard et al. 1971). Reproduced with permission.

As the last topic in our discussion on the structure of the surface layer we consider the results of Monin-Obukhov similarity when applied to other variables the mean velocity and temperature. Let us consider σ_w and σ_ϑ which are the standard deviations of vertical velocity and temperature fluctuations, respectively. According to Monin-Obukhov similarity it should follow that

$$\frac{\sigma_w}{u_*} \equiv \frac{\sqrt{\overline{w^2}}}{u_*} = f_w\left(\frac{z}{L}\right)$$

$$-\frac{\sigma_\vartheta}{\vartheta_*} \equiv -\frac{\sqrt{\overline{\vartheta^2}}}{\vartheta_*} = f_\vartheta\left(\frac{z}{L}\right). \tag{4.8}$$

In figure 14 we have plotted some observations of both σ_w and σ_ϑ in the surface layer. Indeed, the data lie on a single curve, which again can be taken as a confirmation of Monin-Obukhov similarity.

It may seem that Monin-Obukhov similarity is generally valid in the surface layer. However, for some parameters Monin-Obukhov similarity fails and an example is the standard deviation of the horizontal velocity fluctuations σ_u and σ_v. The explanation of this failure will be deferred to section 4.3

Dynamics   We have shown in the previous section that transport processes in the surface layer can be parameterized in terms of K-theory. This means that the components of the Reynolds stress in the boundary-layer equations (3.2) can be closed according to (3.11) and (3.12) and a similar closure is valid for the temperature flux. As a result the time development of the profiles for $\bar{u}$ and $\bar{\vartheta}$, which are forced by the boundary conditions at the surface, can be estimated at height z as

$$\mathcal{T}_m \simeq \frac{z^2}{K_m} \simeq \frac{z}{u_*} \tag{4.9}$$

where in the second step we have used (4.2) with the approximation ϕ_m and $\phi_h \simeq 1$.

However, the ratio z/u_* is also representative of the turbulence time scale in the surface layer because for the turbulence length scale it follows $\ell \simeq z$ and for the velocity scale $v \simeq u_*$. In other words for the surface layer we find

$$\mathcal{T}_m \simeq \mathcal{T}_t.$$

Because also $\mathcal{T}_m \simeq \mathcal{T}_p$ in this case, the turbulence dynamics of the surface layer obeys the first-order balance "production = dissipation".

The consequence of these estimates is that the time scale $\mathcal{T}_m$ is in general small with respect to the forcing time scale $\mathcal{T}_f$, which, as mentioned in the previous subsection, can be estimated as $\mathcal{T}_f \approx 3$ hours for the Coriolis term or $\mathcal{T}_f \approx 6$ hours for the daily cycle whereas with $z \simeq 10\text{m}$ and $u_* \sim 0.3$ m/s it follows that $\mathcal{T}_m \approx 30$ s. As a result of $\mathcal{T}_m \ll \mathcal{T}_f$, the dynamics of the mean profiles in the surface layer can be characterized as quasi-stationary and, for instance, we will expect no influence of rotation nor of the daily cycle on the structure of the surface layer.

Quasi-stationarity means also that the time derivatives in boundary layer equations (3.2) can be neglected and with the other approximation mentioned above this leads to the following equations for the surface layer

$$0 = \frac{\partial}{\partial z}\left(K_m \frac{\partial \bar{u}}{\partial z}\right)$$

$$0 = \frac{\partial}{\partial z}\left(K_h \frac{\partial \bar{\vartheta}}{\partial z}\right) \tag{4.10}$$

where we have chosen the x-axis along the surface stress τ_o so that the velocity component $\bar{v} \equiv 0$. In other words, in the surface layer all profiles are always adjusted to the surface boundary conditions, which are assumed to vary on the time scale $\mathcal{T}_f \gg \mathcal{T}_m$. We again emphasize that this can be true only for a horizontally homogeneous boundary layer. This result is at the same time the justification of our assumption made in the previous section that turbulent fluxes in the surface layer are independent of height.

4.2 Neutral boundary layer

Scaling We now turn to the region of the boundary layer above the surface layer, i.e. $z/h > 0.1$. For the case of the neutral boundary layer, i.e. $\overline{w'\vartheta'}_o \simeq 0$ or buoyancy effects do not play a role, the appropriate scaling parameters are

$$u_* \quad h \quad \text{and} \quad z$$

Before we consider scaling relationships for the neutral boundary layer, one may ask whether this type of boundary layer is relevant for the real atmosphere. For this we have to estimate the magnitude of temperature fluctuations ϑ', for which we can no longer neglect buoyancy. Let us introduce a Richardson number defined as

$$Ri' = \frac{g}{T_o} \frac{\vartheta' \ell}{u_*^2}.$$

It gives the ratio of the acceleration due to a temperature fluctuation ϑ' with respect to the turbulent acceleration u_*^2/ℓ. Let us assume that for $Ri' < 0.1$ buoyancy effects can be neglected. If we take $\ell = 100$ m and $u_* = 0.3$ m/s, we find that this condition is satisfied for $\vartheta' \lesssim 3 \; 10^{-3}$ °C. This is quite a small temperature fluctuation indeed. We must conclude that already for quite moderate temperature fluctuations (say 10^{-2} °C) buoyancy effects in the boundary layer cannot be neglected. Temperature fluctuations of this magnitude are almost always present so the neutral boundary layer must be considered an exceptional case, which will be rarely encountered in the real atmosphere. Nevertheless, a study of the neutral boundary layer is useful because it allows us isolate the influence of shear production on atmospheric turbulence.

Considering the strong limitations under which a neutral boundary layer will occur, it will be also clear that observations will be difficult and scarce. An exception are the experimental data described by Grant (1992). Therefore, most information on this type of boundary layer follows from numerical computations such as large-eddy simulations. An example of the latter is the study by Mason and Thomson (1987). For an investigation in which various large-eddy codes for the neutral boundary layer have been compared, we refer to Andren et al. (1994).

Dynamics For this case of a shear-dominated boundary layer, the turbulent length scale ℓ can be estimated as $0.07h$ (Hinze, 1975). This means that ℓ is considerably smaller than the boundary-layer depth h, which is the distance over which mean variables such as $\bar{u}$ change.

The condition $\ell < h$ is at the same time a necessary condition for the applicability of gradient transfer theory (Tennekes and Lumley, 1972). It thus seems reasonable to estimate vertical transport in a neutral boundary layer in terms of K-theory. Following similar reasoning as used for the surface layer, we then may derive for the time scale of the mean boundary layer structure

$$\mathcal{T}_m \simeq \frac{h^2}{K} \simeq \frac{14h}{u_*}$$

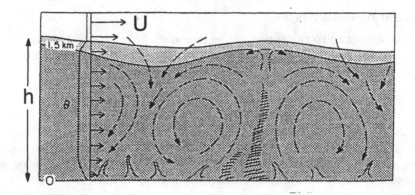

Figure 15. Schematic picture of the convective boundary layer. Reproduced with permission from Wyngaard (1992).

where in the second step we have used $K \simeq u\ell$ and the estimate for ℓ given above. If we take as representative values $h \simeq 500$ m and $u_* \simeq 0.3$ m/s, T_m becomes of the same order of magnitude as T_f, which is estimated as $\sim f^{-1} = 10^4$ s.

Thus, we may conclude that, in contrast to the surface layer, the mean structure of the neutral boundary layer cannot be considered quasi-stationary and, for instance, the influence of the Coriolis effect can not be neglected. In other words, to describe the neutral boundary layer one has to solve the full time-dependent equations for the horizontally homogeneous boundary layer

$$\frac{\partial \bar{u}}{\partial t} - f\,(\bar{v} - V_g) = -\frac{\partial \overline{u'w'}}{\partial z}$$

$$\frac{\partial \bar{v}}{\partial t} + f\,(\bar{u} - U_g) = -\frac{\partial \overline{v'w'}}{\partial z}$$

4.3 Convective boundary layer

Scaling

In figure 15 we show a schematic picture of the convective boundary layer, i.e. $\overline{w'\vartheta'}_o > 0$. The turbulence structure is composed of large eddies, which fill the whole boundary layer. These eddies are directly connected to buoyancy production because they consist mainly of air heated at the surface, which will rise due to buoyancy. These large eddies are known as thermals or plumes. This production process is dominant so that the influence of shear production can be neglected and this means that u_* will no longer play a role in the scaling of the convective boundary layer. Based on this consideration we propose h, $\overline{w'\vartheta'}_o$ and z as characteristic parameters, from which the following scaling parameters can be obtained

$$h \quad w_* \quad \text{and} \quad z$$

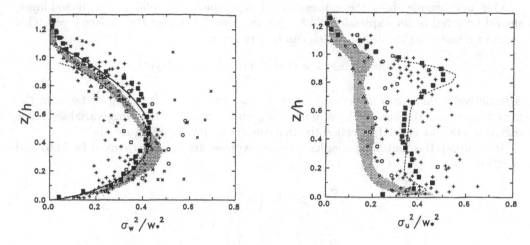

Figure 16. Vertical profiles of the standard deviation of vertical velocity and horizontal velocity fluctuations scaled following mixed layer scaling; the lines denote empirical fits to the observations indicated by the symbols; the shaded area is a compilation of the data obtained from numerical simulations. Reproduced with permission from Hibberd (1994)

where w_* is the convective velocity scale defined as

$$w_* = \left(\frac{g}{T_o} \overline{w'\vartheta'}_o h \right)^{\frac{1}{3}}. \tag{4.11}$$

The scaling of variables in terms of the parameters given above is known as mixed-layer scaling (see figure 8). As example we show in figure 16 the vertical profiles of σ_w and σ_u (for σ_ϑ a similar figure can be shown). According to mixed-layer scaling these variables should obey the following expressions

$$\frac{\sigma_w}{w_*} \equiv \frac{\left(\overline{w^2}\right)^{1/2}}{w_*} = f_w\left(\frac{z}{h}\right)$$

$$\frac{\sigma_u}{w_*} \equiv \frac{\left(\overline{u^2}\right)^{1/2}}{w_*} = f_u\left(\frac{z}{h}\right) \tag{4.12}$$

$$\frac{\sigma_\vartheta}{T_*} \equiv \frac{\left(\overline{\vartheta^2}\right)^{1/2}}{T*} = f_\vartheta\left(\frac{z}{h}\right)$$

where the temperature scale T_* is defined here as $T_* = \overline{w'\vartheta'}_o/w_*$. We find that the data in figure 16 plot reasonably well along a single curve which confirms mixed-layer scaling.

One may wonder how the expressions (4.12), which are valid in the mixed-layer, should be fitted to the expressions (4.8) that we have derived for the surface layer. This requires a matching condition, which can be written as

$$\lim_{z/h \to 0} \text{(mixed layer)} = \lim_{-z/L \to \infty} \text{(surface layer)}$$

Alternatively, this condition can be expressed as: in the matching region between the mixed-layer and the surface layer only $\overline{w'\vartheta'}_o$ and z remain as scaling variables. The region where this is valid, is called the free convection layer (see figure 8).

If we apply this matching condition to the expressions f_w and f_ϑ given by (4.8) and (4.12) we find

$$\frac{\sigma_w}{u_*} \simeq 1.8 \left(-\frac{z}{L}\right)^{1/3},$$

$$\frac{\sigma_\vartheta}{T_*} \simeq 0.9 \left(-\frac{z}{L}\right)^{-1/3}$$

where the numerical coefficients have been obtained by fitting to experimental data. The result of matching is thus an explicit relationship for these standard deviations as a function of height. This follows from the fact that in the free convection layer only a limited number of scaling parameters remain. If we return to figure 14, we see that the free-convection expressions are very well followed by the experimental data.

If we consider the horizontal velocity fluctuations given by (4.12) in the free convection layer, i.e. for $\lim_{z/h \to 0}$, we find

$$\sigma_u \simeq 0.6 w_* \simeq 0.8 \left(-\frac{h}{L}\right)^{1/3} u_*$$

where we have used the data shown in figure 16 to obtain the numerical coefficients. From this expression, it is clear that σ_u does not approach Monin-Obukhov similarity, as we already have mentioned in section 4.1. Here, we find the reason. Even near the surface h remains important as a length scale for the horizontal velocity fluctuations. The background is that even near the surface the large-scale eddies, i.e. the eddies that scale with the boundary-layer height h, are dominant over the surface-layer eddies that scale with z. The process that produces these large-scale horizontal velocity fluctuations near the surface is blocking, which implies that the vertical motion of the large scale eddies is transformed into horizontal motion due to the presence of the surface.

Mixed-layer similarity and free convection scaling are also confirmed by the results of large-eddy simulation (LES). As a matter of fact, w_* as a scaling parameter was first introduced based on results obtained by numerical simulations (Deardorff, 1974). Since then, additional LES results of the convective boundary layer have appeared in the literature, which have led to much insight in the structure of convective turbulence. Examples are Schmidt and Schumann (1988) and Mason (1989). The success of large-eddy simulation is due to that fact that the convective boundary layer is dominated by large-scale eddies that can be easily resolved numerically. For a review comparing several large-eddy codes, we refer to Nieuwstadt et al. (1992).

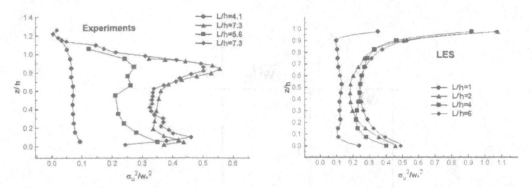

Figure 17. Vertical profiles of the standard deviation of horizontal velocity fluctuations for various aspect ratios, i.e. the horizontal size **L** of the domain with respect to its height with in the left figure the results of laboratory experiments and in the right figure the results obtained from LES. Reproduced with permission from Nieuwstadt (1992)

Nevertheless, even LES is not without problems and an example is illustrated in figure 17 where we show profiles of horizontal velocity fluctuations as a function of the aspect ratio, i.e. the ratio of the horizontal size of the domain with respect to the vertical length scale h. In the left figure of 17, the results of laboratory experiments are shown, while in the right figure the results of LES-simulations are given. It is clear that the experimental values show a much stronger dependence on the aspect ratio than the LES data. The reason for this discrepancy is not clear.

The structure of turbulence in the convective boundary layer has for instance large consequences for the dispersion of pollution inside this boundary layer. The existence of thermal plumes, which transport hot air from the surface directly to the top of the boundary layer, causes pollution emitted near the surface to be quickly transported upward. The concentration pattern in that case shows even an elevated concentration maximum near the top of the boundary layer. For pollution emitted at elevated levels in the convective boundary layer another phenomenon occurs. Because the vertical velocity in the convective boundary layer should be on the average equal to zero (as consequence of the continuity equation) and because of the fact that thermals have a small lateral size but also have a large positive velocity, the air surrounding the thermals must move downward at a vertical speed smaller than the speed inside the thermals. In other words the vertical velocity distribution in a convective boundary layer is skew. the consequence is that for pollution emitted from high stacks in the convective boundary layer, the plume will find itself more frequently in the downward moving air outside the thermals than in the upward moving air inside the thermals. As a result the plume moves downward and strikes the ground leading there to high concentration levels of pollution. For more information on this behaviour of pollution plumes in the convective boundary layer we refer to Weil (1988).

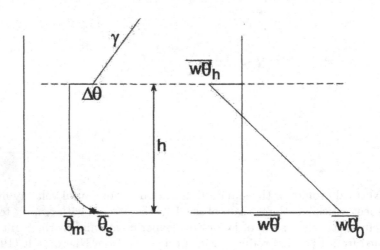

Figure 18. Schematic representation of the mean temperature (left) and the temperature flux (right) profile in a convective boundary layer with $\overline{w'\vartheta'}_o$ the temperature flux at the surface and $\overline{w'\vartheta'}_h$ the entrainment flux at the top of the boundary layer.

Dynamics   We have already mentioned that turbulent eddies in convective conditions fill the whole boundary layer so that a reasonable estimate for the turbulence length scale is $\ell \simeq h$. This means that the condition for K-theory is not satisfied. Moreover, turbulence dynamics is dominated by large-scale vertical transports by these eddies. This implies that the whole boundary layer is strongly coupled so that the time scale of the mean structure will be close to the turbulent time scale: $T_m \simeq T_t$. The T_t may be estimated as h/w_*. With $h \sim 1000$ m and $w_* \simeq 1$ m s^{-1}, we find that $T_t \sim 15$ min. For our choice of the time scale $T_f \simeq f^{-1}$, we find that $T_m \ll T_f$. This leads us to the important conclusion that the mean structure of the convective boundary layer is quasi-stationary and that the effects of rotation may be neglected.

Another fact is that due to the buoyancy production the turbulence intensity in the convective boundary layer is large. As a result one may expect efficient vertical mixing so that over most of the boundary layer the mean variables can be described by a well-mixed profile, which for the average temperature is illustrated in figure 18. Moreover, due to quasi-stationarity, the shape of this well-mixed profile is independent of time although the actual value of the mean temperature $\bar{\vartheta}_m$ must of course increases as a function of time due to the heat input at the surface.

With the time-invariant shape of $\bar{\vartheta}(z)$ as point of departure, we consider the energy conservation equation (3.3), which for our case of a horizontally homogeneous boundary layer becomes

$$\frac{d\bar{\vartheta}}{dt} = -\frac{\partial \overline{w'\vartheta'}}{\partial z}. \tag{4.13}$$

Let us integrate this equation between $0 < z < h$ and $h^- < z < h^+$ with as result

$$\frac{d\bar\vartheta_m}{dt} = \frac{\overline{w'\vartheta'}_o - \overline{w'\vartheta'}_h}{h}$$

$$\Delta\theta \frac{dh}{dt} = -\overline{w'\vartheta'}_h \tag{4.14}$$

where for the second integration we have used that at the top of the boundary layer the temperature exhibits a jump discontinuity $\Delta\theta = \bar\vartheta(h^+) - \bar\vartheta(h^-)$. The $\overline{w'\vartheta'}_h$ is then the temperature flux at $z = h^-$. For $\Delta\theta$ we can derive from the temperature equation (4.13)

$$\frac{d\Delta\theta}{dt} = \gamma \frac{dh}{dt} - \frac{d\bar\vartheta_m}{dt} \tag{4.15}$$

where γ is the temperature gradient above the boundary layer.

The first equation of (4.14) implies that the temperature flux is a linear function of z. Such linear profile is the direct consequence of quasi-stationarity or the time-invariant profile shape mentioned above.

The set of equations (4.14) and (4.15) cannot be solved because there are more unknowns than equations. We have thus again a closure problem. This is solved by expressing $\overline{w\vartheta}_h$ in terms of known variables. By integrating the energy equation (2.14) for quasi-stationary and horizontally homogeneous conditions across the boundary layer and using $\epsilon = w_*^3/h$, we find

$$\overline{w'\vartheta'}_h = -\beta \overline{w'\vartheta'}_o. \tag{4.16}$$

where from experimental data it follows that the constant β is $0.1 - 0.2$. This result can be written alternatively as

$$\frac{dh/dt}{w_*} = \frac{\beta}{Ri_{w*}} \tag{4.17}$$

with $Ri_{w*} = (g/T_o)\Delta\theta h/w_*^2$. For a more detailed discussion on the background of this so-called entrainment assumption, we refer to Garratt (1992, p. 158).

For the case of a constant $\overline{w'\vartheta'}_o$, the set of equations (4.14), (4.15) and (4.16) can be solved explicitly with as result

$$h^2 - h_0^2 = 2\frac{1 + 2\beta}{\gamma} \overline{w'\vartheta'}_o t \tag{4.18}$$

where we have omitted the contribution of a fast decaying transient. The h_0 is the initial height of the boundary layer at $t = 0$.

With this equation and with a given value for $\overline{w'\vartheta'}_o$ we can make an estimate of the boundary-layer height. A representative value for $\overline{w'\vartheta'}_o$ is 0.1 m K s^{-1}. For γ we take 0.01 °C m^{-1}. Equation (4.18) then leads to the result that over a period of 8 hours the convective boundary layer can grow by $\sim$ 900 m. The convective boundary layer is thus in general quite deep and this is consistent with the large turbulence activity in convective conditions.

With the solution of $h = h(t)$, all the parameters for mixed-layer scaling are known so that we can determine the structure of our convective boundary layer at each time instant.

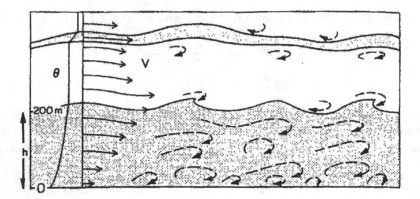

Figure 19. Schematic picture of the stable boundary layer. Reproduced with permission from Wyngaard (1992)

4.4 Stable boundary layer

Scaling   A schematic picture of the stable boundary layer is shown in figure 19. Compared with the convective boundary layer illustrated in figure 15, we see a completely different structure. Most noticeable is the fact that the boundary layer is rather shallow with a height of around 100 m, while we have seen that the convective boundary layer measures around 1000 m. Moreover, we have argued that in a stable boundary layer buoyancy effects oppose vertical motion or alternatively the buoyancy production term in (2.14) is negative. Consequently, turbulent eddies will be small with as result that their length scale[19] $\ell \ll h$.

So in stable conditions shear production remains as the only source of turbulence and it will be clear that turbulence must be scaled in terms of this process. However, the correct scaling velocity in this case is not the surface friction velocity u_*. Because of the fact that $\ell \ll h$, turbulence at each height z will primarily be determined by local processes. In other words, turbulent eddies do not feel the influence of the surface directly. As a result the appropriate scaling parameters become

$$\overline{w'\vartheta'} \quad \tau/\rho_o \quad \text{and} \quad z$$

where $\overline{w'\vartheta'}$ and τ stand for the turbulent temperature and momentum flux at height z. With these parameters and the buoyancy parameter g/T_o we can define a length scale

$$\Lambda = -\frac{(\tau/\rho_o)^{3/2}}{k\frac{g}{T_o}\overline{w'\vartheta'}} \tag{4.19}$$

which is known as the local Obukhov length.

---

[19]This is in particular the case for the vertical length scale of eddies and this leads to small vertical transport in a stable boundary layer. In contrast the horizontal length scale can become quite large.

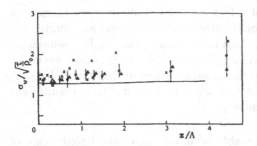 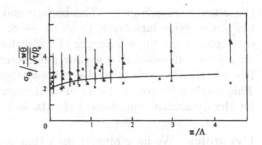

Figure 20. Local scaling applied to the standard deviation of the vertical velocity fluctuations (σ_w) non-dimensionalized with $\sqrt{\tau/\rho_o}$ and the temperature fluctuations (σ_ϑ) non-dimensionalized with $-\overline{w'\vartheta'}/\sqrt{\tau/\rho_o}$; the solid line is a second-order turbulence model; each symbol denotes the average over several observations and the vertical lines give the variation among these observations (Nieuwstadt, 1984a). Reproduced with permission.

Scaling in terms of the parameters mentioned above is known as local scaling (Nieuwstadt, 1984a), which is also indicated as a scaling regime in figure 8. Note that local scaling can be considered as a generalization of Monin-Obukhov similarity of the surface layer discussed in section 4.1 in which the surface stress u_* and Obukhov length L, based on surface fluxes, were used as scaling parameters. As a consequence local scaling matches automatically to Monin-Obukhov similarity in the limit $z \to 0$.

Application of local scaling to the standard deviation of the vertical velocity and temperature fluctuations leads to

$$\frac{\sigma_w}{\sqrt{\tau/\rho_o}} \equiv \sqrt{\frac{\overline{w^2}}{\tau/\rho_o}} = f_w\left(\frac{z}{\Lambda}\right)$$

$$\frac{\sigma_\vartheta}{\overline{w'\vartheta'}/\sqrt{\tau/\rho_o}} \equiv \sqrt{\frac{\overline{\vartheta^2}}{\overline{w'\vartheta'}^2/(\tau/\rho_o)}} = f_\vartheta\left(\frac{z}{\Lambda}\right). \tag{4.20}$$

These expressions are compared with observations in figure 20. Taking into account the large scatter which is customary for measurements in the stable boundary layer, we find reasonable confirmation for our local scaling approach.

Another parameter, to which local scaling can be applied, is the so-called gradient Richardson number Ri defined as

$$Ri = \frac{\frac{g}{T_o}\frac{\partial\bar{\vartheta}}{\partial z}}{\left(\frac{\partial\bar{u}}{\partial z}\right)^2}. \tag{4.21}$$

The Richardson number can be interpreted as a measure of local stability and for instance a theoretical result is that a flow is stable to small perturbations when $Ri > 1/4$. According to local scaling, Ri should be a function only of z/Λ.

When in the local scaling the height z becomes large, the dependence of scaling relationships such as (4.20) becomes independent of height. In figure 8 this is called the z-less scaling regime. The background of z-less scaling is that at large heights in the stable boundary layer, i.e. $z/\Lambda \to \infty$, local turbulence conditions are completely decoupled from direct influence of the surface. As a consequence, the height z should no longer be considered as a scaling parameter. It implies that all locally scaled variables (such as the Richardson number) should approach a constant value in the limit $z/\Lambda \to \infty$. This result will prove to be useful in the next section when we need a closure assumption for the dynamical equations of the stable boundary layer.

Dynamics   We have already seen that in a stable boundary layer the length scale of turbulence ℓ is much smaller than the boundary-layer height h. As we have noted during our treatment of the neutral boundary layer, this means that the necessary condition for the applicability of K-theory is satisfied. In other words, the dynamics of the stable boundary layer is again dominated by diffusion. Similar to what we have done for the neutral boundary layer, we may estimate the time scale of mean structure of the boundary layer as

$$\mathcal{T}_m = \frac{h^2}{K_m}.$$

Let us assume that this mean time scale is proportional to the forcing time scale ($\mathcal{T}_f \simeq f^{-1}$). Next we must estimate the K_m. According to local scaling, discussed in the previous section, the most logical estimate would be $K_m \simeq \sqrt{\tau/\rho_o}\,\Lambda$. However, both variables $\sqrt{\tau/\rho_o}$ and Λ are still unknown functions of height whereas we would prefer to have a constant order of magnitude estimate for K_m. Let us therefore take the values that these variables have in the surface layer: u_* and L so that $K_m \simeq u_* L$. We know that in stable conditions both u_* and L are small. Consequently the K in a stable boundary layer is small or alternatively mixing is slow. This is consistent with our discussion of the effect of stability in the surface layer (see section 4.1).

Substituting the estimate found for K_m and using $\mathcal{T}_m \simeq f^{-1}$, we can use the expression for $\mathcal{T}_m$ given above to estimate the height of the stable boundary layer. The result reads

$$h = c\sqrt{\frac{u_* L}{f}}. \tag{4.22}$$

where c is a proportionality constant, the value of which will be discussed below. This expression is known as the Zilitinkevich height.

One may wonder, under which conditions expression (4.22) for the boundary-layer height is valid. Given our assumption that the time scale $\mathcal{T}_m \simeq \mathcal{T}_f$ we must conclude that the mean structure of the stable boundary layer can not be (quasi)stationary and an equation to describe its time evolution is needed. Such an equation for the boundary-layer height h can be found by integrating the equations that describe the mean structure as a function of time over the height of the boundary layer. The result reads (Nieuwstadt and Tennekes, 1981)

$$\frac{dh}{dt} = -\frac{1}{T_r}(h - h_{eq}). \tag{4.23}$$

We thus find that the stable boundary-layer height follows a relaxation equation, which means that h as function of time approaches an equilibrium height h_{eq}. It turns out that this equilibrium height is equal to the Zilitinkevich expression (4.22). However, we can also learn from equation (4.23) another fact about the stable boundary layer. Namely, the time scale T_r is given by

$$T_r \sim \left(-\frac{1}{\Delta\Theta}\frac{dT_o}{dt}\right)^{-1}$$

where dT_o/dt is the rate of temperature change at the surface and $\Delta\Theta$ is the mean temperature difference across the boundary layer. Typical values for dT_o/dt and $\Delta\Theta$ are 1 °C/hour and 5 °C, respectively. This leads to $T_r \simeq 5$ hours. In other words, the development of the stable boundary layer is slow, which is consistent with the small mixing mentioned above.

Let us now consider the structure of the stable boundary layer in somewhat more detail. To this end restrict ourselves to a so-called quasi-stationary boundary layer. For the case of quasi-stationarity all turbulent quantities are independent of time, from which it follows that

$$\frac{\partial}{\partial t}\left(\frac{\partial \bar{u}}{\partial z}\right) = \frac{\partial}{\partial t}\left(\frac{\partial \bar{v}}{\partial z}\right) = \frac{\partial}{\partial t}\left(\frac{\partial \bar{\vartheta}}{\partial z}\right) = 0$$

It is clear that this will be a quite exceptional case because we have seen above that in general the stable boundary layer is non-stationary. Furthermore, we shall assume that turbulence in our stable boundary layer can everywhere be described in terms of z-less scaling, i.e. we neglect everywhere the direct influence of the surface. We have seen that for the Richardson number (4.21) this implies

$$Ri = Ri_{cr} \simeq 0.2 \tag{4.24}$$

where the numerical value follows from experiments.

Applying the results mentioned above for the mean velocity and temperature gradients to (3.9) and (3.10) and assuming that the geostrophic wind is independent of height, we find that the following equations apply to the quasi-stationary stable boundary layer

$$-f\frac{\partial \bar{v}}{\partial z} = \frac{\partial^2 \overline{u'w'}}{\partial z^2} \tag{4.25}$$

$$f\frac{\partial \bar{u}}{\partial z} = \frac{\partial^2 \overline{v'w'}}{\partial z^2} \tag{4.26}$$

$$0 = \frac{\partial^2 \overline{w'\vartheta'}}{\partial z^2}. \tag{4.27}$$

Together with (4.24) as closure assumption this set of equations allow a closed-form solution, the details of which are described in (Nieuwstadt, 1984a) and (Garratt, 1992; pp 165-170). One of the results, which is immediately obvious from (4.27) is that the temperature flux should follow a linear profile as a function of height. In addition it can be derived that the absolute value of the Reynolds stress $|\tau|$ is given by the following profile

$$|\tau| \equiv \sqrt{\tau_x^2 + \tau_y^2} = \rho_o u_*^2 \left(1 - \frac{z}{h}\right)^{\frac{3}{2}} \tag{4.28}$$

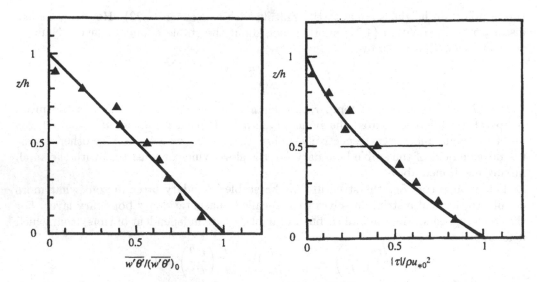

Figure 21. Vertical profiles of the temperature flux $\overline{w'\vartheta'}$ and the momentum flux τ in the quasi-stationary stable boundary layer; the symbols denote an average over a number of experiments; the variation between the experimental results is given by the horizontal line; the solid lines are the closed-form expression discussed in the text.

where we remind that $u_* = \sqrt{|\tau|_o/\rho_o}$. These results for the temperature flux and stress profile are compared with experiments in figure 21 and although the assumption of a quasi-stationary boundary layer is quite restrictive, the agreement of the linear temperature flux profile and (4.28) with the experimental data is quite reasonable.

The closed-form solution for the gradients of the mean velocity and temperature profile are found to be (Nieuwstadt, 1984a) and (Garratt, 1992; pp 165-170)

$$
\begin{pmatrix}
\partial \bar{u}/\partial z \\
\partial \bar{v}/\partial z \\
\partial \bar{\vartheta}/\partial z
\end{pmatrix}
= \frac{1}{k\,\mathrm{Ri}_f} \frac{u_*}{L}
\begin{bmatrix}
(1 - z/h)^{-\frac{1}{2}} \cos\left\{ \frac{1}{\sqrt{3}} (1 - z/h)^{-\frac{1}{2}} \right\} \\
-(1 - z/h)^{-\frac{1}{2}} \sin\left\{ \frac{1}{\sqrt{3}} (1 - z/h)^{-\frac{1}{2}} \right\} \\
\frac{\mathrm{Ri}}{k\,\mathrm{Ri}_f^2} (1 - z/h)^{-1}
\end{bmatrix}
$$

where the flux-Richarson number Ri_f is defined as

$$
\mathrm{Ri}_f = \frac{g/T_o\,\overline{w'\vartheta'}}{\overline{u'w'}\,\partial \bar{u}/\partial z + \overline{v'w'}\,\partial \bar{v}/\partial z}. \tag{4.29}
$$

These equations can be integrated further to produce solutions for the mean velocity and temperature profile in the quasi-stationary stable boundary layer. The solution for the velocity profile leads to the angle α_g between the geostrophic wind at the top of the boundary layer and the surface stress equal to $\alpha_g = \pi/3$, which is larger than the value

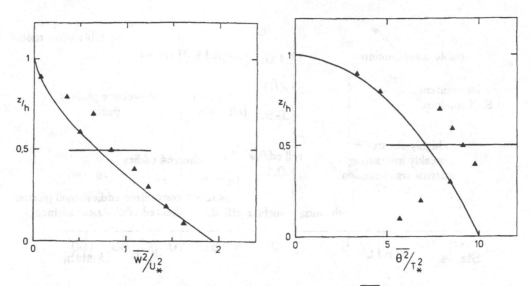

Figure 22. Vertical profiles of the variance of the vertical $\overline{w'^2}$ and the variance of the temperature fluctuations $\overline{\vartheta'^2}$ in the quasi-stationary stable boundary layer; the symbols denote an average over a number of experiments; the variation between the experimental results is given by the horizontal line; the solid lines are the closed-form expression discussed in the text.

$\pi/4$, which we have found in section 3.1 for the Ekman solution. Furthermore, we find

$$\frac{G}{u_*} = \frac{1}{k\,\mathrm{Ri_f}}\frac{h}{L} \tag{4.30}$$

which can be interpreted as a resistance law. For the mean temperature profile the closed form solution reads

$$\frac{\bar{\vartheta} - \vartheta_0}{\vartheta_*} = -\frac{\mathrm{Ri}}{k\,\mathrm{Ri}_f^2}\frac{h}{L}\ln\left(1 - \frac{z}{h}\right) \tag{4.31}$$

which leads to a logarithmic singularity at the top of the boundary layer.

Together with the application of z-less scaling, the closed-form solutions given above lead to the following profiles for the velocity and temperature fluctuations

$$\sigma_u, \sigma_v, \sigma_w \;\simeq\; \left(\frac{\tau}{\rho}\right)^{\frac{1}{2}} = c_u, c_v, c_w \left(1 - \frac{z}{h}\right)^{\frac{3}{4}} \tag{4.32}$$

$$\sigma_\vartheta \;\simeq\; \frac{\overline{w'\vartheta'}}{(\tau/\rho)^{1/2}} = c_\vartheta \left(1 - \frac{z}{h}\right)^{\frac{1}{4}}. \tag{4.33}$$

These profiles with for the constants $c_w = 1.4$ and $c_\vartheta = 3.1$ are compared with experimental data in figure 22. Again the agreement between the relationships (4.32) and

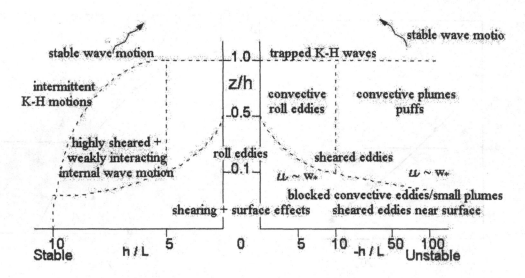

Figure 23. Coherent structures for the various scaling regimes of the atmospheric boundary layer.

(4.33) is quite reasonable in view of the restrictive assumption of a quasi-stationary stable boundary layer.

An additional result, which follows from the closed form solution of (4.25) - (4.27), is that the height of the quasi-stationary stable boundary layer, which is equivalent to the equilibrium height mentioned in connection with (4.23), is given by (4.22) with the constant c equal to

$$c = \left(\sqrt{3}\, k\, Ri_{cr} \right)^{1/2} \simeq 0.37.$$

where the numerical value follows from subsituting $Ri_{cr} = 0.2$ and $k = 0.4$. The values for the stable boundary-layer height, which follow from expression (4.22) with $c = 0.37$ are found to be in good agreement with observed boundary-layer heights (Nieuwstadt, 1984b). To give an example with some representative values for u_* and L, say 0.2 m s^{-1} and 25 m, respectively, we obtain from (4.22) that $h \approx 80$ m, which is close to the estimate $h \simeq 100$m mentioned before. We may conclude that compared to the convective boundary layer, discussed in section 4.3, the stable boundary layer is indeed quite shallow.

5 Additional topics

5.1 Coherent structures

Until now we have discussed the structure and dynamics of the boundary layer in terms of statistics such as the mean velocity and temperature and the variances and correlations of the fluctuations. However, turbulence is not a random process, which can only be described statistically. On the contrary, turbulent flow, especially with respect to its large scale structure, consists of recognizable and reproducible structures, like the large-scale thermal plumes, which we have already noticed in the convective boundary layer. These characteristic flow patterns are in general denoted as coherent structures. In figure 23 we have indicated the various types of coherent structures organized according to the scaling regimes of figure 8.

In the surface layer the structures, which we may also denote as eddies, are dominated by the presence of the surface, which results in shear and the effect of blocking. Above the surface layer the type of structure depends on whether the boundary is stable or unstable. In near-neutral conditions the roll structures[20] may develop, which we have already encountered in section 3.1. In the free atmosphere above the boundary layer, where conditions are in general stable, gravity waves can be found, which may be generated by e.g. turbulent eddies impinging on the top of the boundary layer. At very strong stability turbulence in the boundary can not survive and becomes intermittent. This means that the flow is laminar with isolated patches of turbulence. The turbulence in these patches can for instance be generated by strong shear through the process of Kelvin-Helmholtz instability. For more details on the coherent structures in the boundary layer we may refer to Nieuwstadt and Hunt (2002).

5.2 Clouds

Next we turn to the case that the liquid phase[21] of water is also present in the boundary layer usually in the form of clouds. The specific liquid water content of the air is given by q_l, which is the mass of liquid water per unit mass of moist air. The definition of the virtual temperature now becomes

$$T_v = T(1 + 0.61q_v - q_l). \tag{5.1}$$

where we remind that q_v is the specific humidity defined as the mass of water vapour per unit mass of moist air

The presence of clouds in the boundary layer leads to considerable complications. The reason for this is that, besides the turbulent fluxes such as the momentum and heat flux introduced at the surface, additional processes such as radiation and phase changes play an important role, because they produce local sources of heating (or cooling) within the interior of the boundary layer and in view of the buoyancy production term in (2.14) can greatly influence the turbulent structure and dynamics and consequently the development of the boundary layer.

In view of the fact that heat exchange due phase change plays a role during cloud formation, we must introduce a new type of conservative variable, which, as we have mentioned in section 2, is a variable that stays constant under vertical and isentropic exchange. Here we shall adopt the so-called equivalent temperature, which is defined as

$$\vartheta_e = \vartheta_v + \frac{L_v}{c_p}q_v \tag{5.2}$$

---

[20] Such roll structures may also develop in the convective boundary layer when the wind velocity is sufficiently strong. In this case the roll structures are sometimes visible as cloud streets.

[21] The ice phase of water will not be considered here

where L_v is heat of water vaporization. Another conservative variable needed to describe clouds in the boundary layer is the total water content q_t which is defined as

$$q_t = q_l + q_v,$$ (5.3)

which is conservative unless water dissappears from the boundary layer, e.g. due to rain.

We can recognize two types of boundary-layer clouds, which are directly connected with turbulent processes in the boundary layer. These are cumulus and stratocumulus clouds. Cumulus clouds are invidual patches of condensed water vapour, which appear in a convective boundary layer when the boundary-layer height is larger than the so-called lifting condensation level (LCL). The lifting condensation level is defined as the height where the temperature of a flow parcel, when moving adiabatically upwards from the surface, becomes smaller than the saturation temperature. In most case the LCL is a well-defined height in the boundary layer, which explains the flat underside of cumulus clouds. As soon as condensation occurs the temperature gradient changes from the dry adiabat γ_d to the moist saturated adiabat γ_w given by (Iribarne and Godson, 1981) as

$$\gamma_s \simeq \gamma_d \frac{1 + \frac{L_v q_s}{R_d T}}{1 + \frac{\epsilon L_v^2 q_s}{c_p R_D T^2}}$$ (5.4)

with q_s the saturated specific humidity at temperature T and ϵ the ratio of the molecular weights of water and dry air. The γ_w is larger than γ_d so that in a cloud the absolute temperature does not decrease as fast with height as in the dry atmosphere, which of course is due to the heat release as result of condensation. This heat release also leads to the production of additional turbulence due to buoyancy production, which results in the typical bulgy shapes on the upper side of cumulus clouds.

We have seen that at the top of the convective boundary layer we usually find a strong inversion or a positive temperature jump $\Delta\theta$. We can now distinghuish two cases. The first case is when the temperature in the cumulus cloud, which follows the moist saturated adiabat, becomes larger than $\vartheta_m + \Delta\theta$, which is the temperature just above the top of the boundary layer. This means that the cumulus cloud is statically unstable with respect to the conditions above the boundary layer. Due to this static instability, the cumulus cloud will rise into the free atmosphere above the boundary layer. When the conditions are suitable, (e.g. when the temperature gradient in the free atmosphere stays below the moist saturated adiabat) a large cumulonimbus cloud may develop and when enough moisture is present, such a cloud can sometimes rise up to tropopause and even into the statosphere.

In the other case when the temperature in the cumulus cloud at the top of the boundary layer is smaller than $\vartheta_m + \Delta\theta$, the cumulus cannot penetrate the inversion and the growth of the cumulus cloud is limited to the boundary-layer height. However, the cloud may grow in a lateral direction until a closed cloud deck is formed. Such a closed cloud deck, of which the top coincides with the boundary-layer height is called stratocumulus. In the remainder of this section the discussion will be restricted to the properties of such a closed stratocumulus deck.

The general picture for such a closed stratocumulus deck is that longwave cooling at cloud top, which occurs in the few tens of meters adjacent to the cloud top, leads to cold air, which due to buoyancy will sink in the form of plumes of cold (moist) air through the cloud layer (Nicholls, 1989). This cold downward moving air results in a positive vertical buoyancy flux throughout the cloud layer and thus according to (2.14) it produces turbulence. As a result the cloud layer is maintained in a well-mixed state. Besides the longwave cooling, the surface fluxes can also generate turbulence in the boundary layer, in particular during daytime or under strong wind conditions. If these fluxes are strong enough, a well-mixed state can be maintained across the whole boundary layer.

Similarly, as in the convective boundary layer, the turbulence produced by the long-wave radiation at the top of the cloud layer causes entrainment, i.e. air is mixed from above into the cloudy boundary layer. The entrained air is typically warmer and dryer than the boundary-layer air. Thus, entrainment is a mechanism for a possible thinning and dissolving of the cloud layer.

In addition, other processes play a role in the energy budget of the cloud. At cloud base, the longwave radiation typically causes a slight warming, since the ground temperature is usually a few degrees higher than the cloud temperature. Furthermore, due to condensation of water vapour there is also a release of latent heat, which also leads to warming of the cloud layer near its base, which we have also called the LCL. Both processes can be also considered as an additional source for buoyant production of turbulence in the cloud layer. Another consequence of these processes at cloud base may be that the temperature at cloud base becomes higher than the temperature in the clear air just below cloud base. A similar condition may also occur during the daytime when the development of the cloud layer will be influenced by the absorption of solar radiation. Because the shortwave heating[22]extends much deeper into the cloud layer than the long-wave cooling, it may also cause a higher temperature of the cloud base with respect to the sub-cloud air. In other words an inversion develops at cloud base, which as a result of the strong stability or alternatively as a result of $\overline{w'\theta'}_{LCL} < 0$ destroys turbulence at this position. When this occurs the cloud layer is said to be decoupled from the sub-cloud boundary layer. As a result turbulent transport becomes weak near the LCL so that moisture is prevented from being transported upwards from the sea surface into the cloud layer. This means that the cloud is cut off from the moisture input from the surface, and as a consequence a rapid thinning of the cloud layer occurs resulting in break up or at least in a marked diurnal cycle.

The liquid water content within stratocumulus increases roughly linearly with height at a value slightly below the adiabatic value. Aircraft observations show that there is usually little variation in droplet concentration with height. Therefore, the deeper the cloud layer, the larger the drops that form within it, and there is a critical depth at which drizzle forms. Cloud microphysics reveals two relevant facts: the droplets with a radius less than about 20 μm grow only by condensation, and the droplets with radius larger than about 20 μm grow only by coalescence (provided they are surrounded by droplets of different sizes). Therefore, to produce drizzle, a few droplets must grow to at least 20 μm radius by condensation.

From the discussion given above we must conclude that many processes play a role in

---

[22]The shortwave heating may even compensate the longwave cooling resulting in dissolving of the cloud layer when the temperature rises above the saturation temperature.

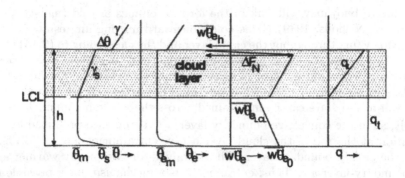

Figure 24. Schematic representation of the profile for the mean temperature, the mean (equivalent) temperature, the (equivalent) temperature flux and the specific humidity in a cloudy boundary layer with $\overline{w'\vartheta'}_{e_o}$ the temperature flux at the surface and $\overline{w'\vartheta'}_{e_h}$ the entrainment flux at the top of the boundary layer; the ΔF_N is the net long-wave radiation jump at the cloud top.

a cloudy boundary layer. As a result it is impossible to define a limited number of scaling parameters, which can be used to describe the structure of the cloudy boundary layer. A simple scaling approach such as we have used for the prototypes of the dry boundary layer, is therefore not possible.

Dynamics   If the cloud layer is not decoupled from the sub-cloud layer, we can assume similarly as for the convective boundary layer that the mean (conservative) variables ϑ_e and q_t are well mixed. This leads to a structure of the cloudy boundary layer as sketched in figure 24. Apart from the variables that we already introduced with respect to the dry convective boundary layer such as the surface temperature flux and the entrainment flux and (note that in the case of the cloudy boundary layer these fluxes should be expressed in terms of the equivalent temperature, i.e. $\overline{w'\vartheta'}_{e_o}$ and $\overline{w'\vartheta'}_{e_h}$ respectively), we also encounter the net long-wave radiation flux F_N and the temperature flux at the bottom of the cloud layer $\overline{w'\vartheta'}_{e_{LCL}}$. Again due to quasi-stationarity, the shape of the well-mixed profile for the equivalent temperature stays independent of time.

Let us first concentrate on the entrainment by which the boundary layer can grow. The entrainment velocity (w_e) in analogy with (4.14) is defined as

$$w_e = -\frac{\overline{w'\vartheta'}_{vh}}{\Delta\theta}.$$

The growth of the boundary layer dh/dt can then be expressed as

$$\frac{dh}{dt} = w_e + w_h$$

where w_h is a velocity, which is determined for instance by large-scale meteorological processes. For instance, in a high pressure system we have due to Ekman pumping a

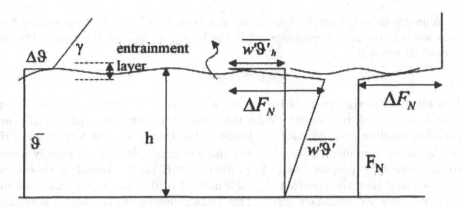

Figure 25. Schematic representation of the profiles for the mean temperature, the temperature flux and the net long-wave radiation in a smoke layer.

downward velocity ($w_h < 0$), which is also called subsidence. In that case the cloudy boundary layer can reach a stationary height when $w_e = -w_h$. Such condition occurs in the extensive marine stratocumulus clouds, which are a persistent feature of the eastern parts of the major ocean basins. Given the subsidence velocity, the entrainment velocity will determine the final depth of the boundary layer.

To simplify our discussion with respect to the dynamics of a cloudy boundary layer we shall limit ourselves to a case without any effect of phase change. Moreover the cloud fills the complete boundary layer from the surface to the top and the only process, which produces turbulence is the long-wave radiation, which is concentrated at the top of the boundary layer. This boundary is sometimes denoted as a smoke layer and its structure is sketched in figure 25.

The equations, which describe the evolution of this smoke layer, read

$$\frac{d\bar{\vartheta}}{dt} = \frac{-\overline{w'\vartheta'}_h}{h} - (1-r)\frac{\Delta F_N}{\rho_0 c_p h} \tag{5.5}$$

$$\Delta\bar{\vartheta}\left(\frac{dh}{dt} - w_h\right) = -\overline{w'\vartheta'}_h + r\frac{\Delta F_N}{\rho_0 c_p} \tag{5.6}$$

$$\frac{d\Delta\bar{\vartheta}}{dt} = \gamma\frac{dh}{dt} - \frac{d\bar{\vartheta}}{dt} \tag{5.7}$$

where r is the fraction of net longwave radiation that occurs within the entrainment layer and $1-r$ the fraction of net longwave radiation that occurs above the entrainment layer.

The equations (5.5) - (5.7) are analogous to the equations (4.13) - (4.14) for the dry convective boundary layer in that their solution also requires a closure assumption. However, such a closure assumption is more complicated in this case because instead of upward moving warm thermals produced at the surface as in the dry convective boundary layer we have now downward moving cold thermals produced at the top of the boundary layer, i.e. at the same location where the entrainment takes place. Therefore, we expect the entrainment to be less efficient as for the dry boundary layer. Several approaches for a closure assumption have been proposed in the literature and for the case of the smoke layer these all result in

$$\overline{w'\vartheta'}_h = -\beta\overline{w'\vartheta'}_{max} \tag{5.8}$$

which is identical to the result (4.16) for the dry convective boundary layer, but with a constant β that will be smaller. With this result the equations (5.5) - (5.7) can be integrated to produce a solution for the height of the smoke layer as a function of time.

For the cloudy boundary layer, in which phase changes play a role, usually a similar closure relationship is proposed as (5.8) or alternatively (4.17). Based on observations, however, we find that the constant β is 2.0 instead of 0.2, the value which we found for the dry convective boundary layer. This large difference in the scaled entrainment velocity for the cloud-topped boundary layer is a consequence of the fact that we have used $\Delta\Theta_v$ to denote the strength of the inversion in the definition of Ri_{w*}. However, $\Delta\Theta_v$ does not take into account the effect of evaporative cooling on the entrainment process and therefore another parameter has to be added to the list of parameters to scale the entrainment velocity. The entrainment into a cloudy boundary layer is still an area of active research.

5.3   Non-stationary and inhomogeneous boundary layers

Until now we have limited our discussion to (quasi)stationary conditions in the sense that the time-scale of turbulence $T_t = \ell/v$ is smaller than or equal to time scale of turbulence production T_p, where the latter is usually taken to be proportional to T_m, the time scale of mean flow development. This restriction means that the turbulence is always in equilibrium, which implies that turbulence production is in balance with dissipation or alternatively that the time derivative in turbulence kinetic energy equation (2.14) is equal to zero. However, there are also situations when this condition does not apply. An example is the atmospheric boundary-layer around sunset when the surface heat input quickly goes to zero at a time-scale smaller than the T_t. Nieuwstadt and Brost (1986) have investigated this case for the simplified conditions that the surface heat flux in a convective boundary layer is turned off instantaneously. They found that turbulence decays according to a power law with a time scale h/w_* where the w_* was the convective time scale of the initial convective boundary layer.

Another example of a non-stationary response of the boundary layer is when the boundary layer is advected over a sudden change in surface boundary conditions, such as a change in roughness or a change in the surface heat input. An example is for instance a sea-land transition where the boundary layer moves from a very smooth surface to a rough surface as illustrated in figure 26. Over the land surface an internal boundary layer (IBL) develops, in which the boundary layer adjusts to the new surface conditions. This internal boundary layer must follow from a solution of the full boundary-layer equations (3.2) where for the closure of the turbulent Reynolds stresses it has to be also taken into account that turbulence may not be in equilibrium. The height of this internal boundary layer, h_i, is usually defined as the height where the wind velocity or the turbulent stress is close (say 99%) to its upstream value. Laboratory and field experiments give the

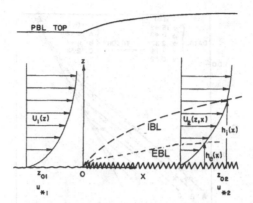

Figure 26. Schematic representation of transition of the boundary layer from a smooth to a rough surface with h_i the height of the internal boundary layer (IBL) and (EBL)the equilibrium boundary layer. Reproduced with permission from Arya (1988).

following result for h_i

$$\frac{h_i}{z_{02}} = a_i \left(\frac{x}{z_{02}}\right)^{0.8}$$

with for the constant a_i a value between 0.35 and 0.75 and where x is the distance from transition. This means that for a roughness of $z_{02} = 0.1$ m and an upstream boundary-layer height of ~ 1000m that it takes about 1.5 km for the internal boundary-layer to fill the whole upstream boundary layer. The structure in this internal boundary layer can be roughly described as being for its lower part close to the boundary layer above the rough surface with roughness z_{02} and being close to the boundary layer above a smooth surface in its upper part with these parts developing as function of x. In other words the internal boundary layer is not horizontally homogeneous and a typical development of the velocity profile as a function of distance is illustrated in figure 27.

Inside the internal boundary layer one can distinguish another layer called the equilibrium boundary layer (EBL). Within this equilibrium layer the flow conditions are fully adjusted to the lower surface, i.e. in our case to a surface with a roughness z_{02}. So within the EBL the conditions can taken to be representative again for a horizontally homogeneous boundary layer. The height of this equilibrium layer is approximately $h_e \simeq 0.1 h_i$ or alternatively

$$h_e \simeq \frac{1}{100} x \sim \frac{1}{200} x.$$

So in order to have an equilibrium layer of at least 10 m high we need a fetch (i.e. the distance between the transition and the point where we consider the equilibrium layer) of at least 1 km. This value should be taken into account when one for instance wants to instal an instrument that should measure properties representative of a particular surface. One needs in that case to have a sufficiently long fetch because otherwise the measurements will be contaminated with upstream information not representative of the surface where the measurement is taken.

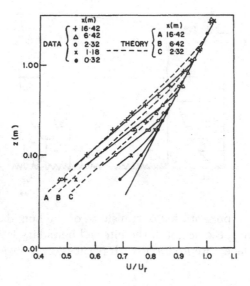

Figure 27. Velocity profiles in a boundary-layer transition from smooth surface ($z_{01} = 0.02$mm) to a rough surface ($z_{01} = 2.5$mm) as obtained from experiments and theory with as reference velocity the wind speed of the upstream boundary layer at 2 m. Reproduced with permission from Arya (1988).

The sea-land transition is also representative for a change in surface heat flux: conditions over sea being mostly neutral or slight stable and the conditions over land being (during the day time) frequently convective. A schematic illustration of this case is given in figure 28. If we assume that the air above the sea is stable with a (potential) temperature gradient γ, a convective boundary layer develops into the stable layer starting from the coast line. The equations for this development are equal to (4.14) and (4.15) with the time derivative replaced by a spatial derivative leading to

$$\bar{u}\frac{d\bar{\vartheta}_m}{dx} = \frac{\overline{w'\vartheta'}_o - \overline{w'\vartheta'}_h}{h}$$

$$\bar{u}\Delta\theta\frac{dh}{dx} = -\overline{w'\vartheta'}_h$$

$$\bar{u}\frac{d\Delta\theta}{dx} = \gamma\frac{dh}{dt} - \frac{d\bar{\vartheta}_m}{dt}$$

where $\bar{u}$ is the mean flow speed in the internal boundary layer, which is taken independent of height and where x is the distance from the coast line. With a closure relation for $\overline{w'\vartheta'}_h$ as given by (4.16) the solution of this set of equations becomes

$$h_i = \sqrt{2\frac{(1+2\beta)\,\overline{w'\vartheta'}_o}{\gamma\bar{u}}x}.$$

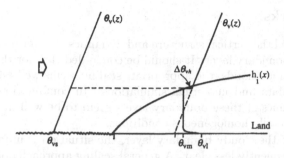

Figure 28. Schematic representation of transition of the boundary layer from a surface with zero heatflux to a surface with a positive heat flux where h_i is the height of the internal boundary layer (IBL). Reproduced with permission from Arya (1988).

With a value for $\overline{w'\vartheta'}_o = 0.1$ m K s^{-1}, for $\gamma = 0.01$ °C m^{-1}, $\beta = 0.2$ and $\bar{u} = 5$ m s^{-1} we find that the internal boundary layer has risen after 1 km to a height of ~ 250 m and has reached the representative height for a convective boundary layer of about 1 km after a fetch of about 18 km.

Such internal boundary layer developement plays for instance an important role in pollution problems where stacks have been built near a shoreline. During day-time when the wind come from the water we have the situation as described above. The stack, when sufficiently high, then emits a plume of pollution into the stable air above the internal boundary layer and this plume is carried along land inward with the wind. At a certain distance, when the internal boundary layer has grown to the height where the plume is located, the pollution is mixed quickly downward to the ground by the strong turbulence in the convective boundary layer. This condition, which can also occur in the morning when the convective boundary grows as a function of time to heights where during the night pollution plumes may have been emitted, is called fumigation.

We have not considered the cases rough-to-smooth and surface heat flux to no surface heat flux, or alternatively the advection of the boundary layer from land to sea. However, similar considerations apply here as we have used for the sea-land transition. The difference is that the turbulence above a sea surface is in general smaller than above a land surface due small roughness length of the water surface and also due to the small heat flux. As a result the growth of the internal boundary layer is in general smaller than what we have found over the land surface. An exception is the case when cold air produced over a land is advected over a warm sea, a situation which may occur during winter time. In that case a convective boundary layer is formed over sea with usually strong cloud formation, which frequently takes the shape of roll structures aligned with the wind.

6 Final remarks

We have considered the vertical structure and dynamics of several so-called prototypes of the atmospheric boundary layer. It should be concluded that for the dry or clear boundary layer, i.e. without clouds, the appropriate scaling seems well established and verified by experimental data and numerical simulations. In combination with this successful scaling, the dynamics of these boundary layers seem to be well understood, at least for the case of horizontally homogeneous conditions.

For the case of the cloudy boundary layer, the situation is unfortunately more complicated and consequently less clear. A general scaling approach does not exist primarily because in comparison with the dry boundary layer more processes play a role. As a result, both the vertical structure and the dynamics of the cloudy boundary layer are not completely understood under all circumstances.

A short discussion has also been presented for the case when the boundary layer deviates from the ideal condition of horizontally homogeniety. In particular the transition from a smooth to a rough surface was considered and the transition from a surface with a zero surface heat flux to a surface with a positive heatlflux. These cases are no more than a few examples of the more general case of a non-stationary and non-homogeneous boundary layer, from which it should be clear that no treatment for these boundary layers exists, which can be applied to all circumstances and conditions.

7 Bibliography

Andren A., Brown A., Graf J., Mason P.J., Moeng C.-H., Nieuwstadt F.T.M. and Schumann U. 1994 Large eddy simulation of a neutrally stratified boundary layer: a comparison of four computer codes. *Q. J. R. Meteorol. Soc.* 120, 1457-1484.

Arya, S.P. 1988 *Introduction to micrometeorology.* Vol 42. International Geophysics Series, Academic Press, New York, U.S.A.

Deardorff J.W. 1974 Three dimensional numerical study of turbulence in an entraining mixed layer *Bound. Layer Meteor.* 7, 199-226.

Durbin P.A. and Petterson B.A. 2000 *Statistical theory and Modeling for Turbulent Flows.* Wiley & Sons Ltd., Chichester, U.K..

Etling, D. and Brown R.A. 1999 Roll vortices in the planetary boundary layer: a review *Bound. Layer Meteor.* 65, 215-248.

Garratt, J.R. 1992 *The atmospheric boundary layer.* Cambridge atmospheric and space science series, Cambridge University Press, Cambridge, England.

Grant A.L.M. 1992 The structure of turbulence in the near-neutral atmospheric boundary layer. *J. Atmos. Sci.* 49, 226–239.

Fleagle R.G. and Businger, J.A. 1980 *An introduction to Atmospheric Physics.* Academic Press, Orlando, Fla., U.S.A.

Folz, A.B. 1970 *An experimental study of teh near-surface turbulence in the atmo-*

spheric boundary layer. Thesis University of Maryland, U.S.A.

Holtslag A.A.M and Nieuwstadt, F.T.M. 1986 Scaling the atmospheric boundary layer. *Bound.-Layer Meteorol.* 36, 201–209.

Hibberd M.F. and Sawford B.L. 1994 A saline laboratory model of the planetary convective boundary layer *Bound.-Layer Meteorol.* 67, 229-250.

Hinze O. 1975 *Turbulence.* Mc Graw-Hill, New York, U.S.A.

Kaimal, J.C. and Finnigan, J.J. 1994 *Atmospheric Boundary Layer Flows; their structure and measurement.* Oxford University Press, Oxford, England.

Launder, B.E. and Sandham, N.D. (eds.) 2002 *Closure strategies for turbulent and transitional flows.* Cambridge University Press, Cambridge, U.K..

Mason P. 1989 Large eddy simulation of the convective atmospheric boundary layer. *J. Atmos. Sci.* 46, 1492–1516.

Mason P.J. and Thomson D.J. 1987 Large-eddy simulation of the neutral-static-stability planetary boundary layer. *Q. J. R. Meteorol. Soc.* 113, 413–443.

Panofsky H.A. and Dutton, S.P. 1984 *Atmospheric Turbulence; Models and Methods for Engineering Applications.* John Wiley & Sons, New York..

Nicholls, S. 1989 The structure of radiatively driven convection in stratocumulus. *Quart. J. Roy .Met. Soc.* 115, 487–511.

Nieuwstadt F.T.M. 1984a The turbulent structure of the stable boundary layer. *J. Atmos. Sci.* 41, 2202–2216.

Nieuwstadt F.T.M. 1984b Some aspects of the turbulent stable boundary layer. *Bound. Layer Meteorol.* 30, 31–54.

Nieuwstadt F.T.M. and Tennekes H. 1981 A rate equation for the nocturnal boundary-layer height. *J. Atmos Sci.* 38, 1418–1428.

Nieuwstadt F.T.M, Mason P.J., Moeng C.-H. and Schumann U. 1992 Large-eddy simulation of the convective boundary layer: a comparison of four computer codes. In *Turbulent Shear Flows 8*, Springer-Verlag, Berlin, pp. 343–367.

Nieuwstadt F.T.M. and Hunt J.C.R. 2002 Boundary Layers: coherent structures. Encyclopedia of Atmospheric Sciences (J.R. Holton, J.A. Pyle and J.A. Curry, eds) Elsevier Sciences, pp. 228–233

Pope S.B. 2000 *Turbulent flows.* Cambridge University Press, Cambridge U.K..

Raupach, M.R., Antonia R.A. and Rajagopalan S. 1988 Rough-wall turbulent boundary layers. *Appl. Mech. Rev.* 44, 1–25.

Schmidt H. and Schumann U. 1988 Coherent structure of the convective boundary layer derived from large eddy simulation. *J. Fluid Mech.* 200, 511–562.

Sorbjan, Z. 1989 *Structure of the atmospheric boundary layer.* Prentice Hall, Engle-Wood Cliffs, N.J. 07632.

Stull, R. 1988 *An introduction to boundary-layer meteorology.* Atmospheric Science

Library, Kluwer Academic Publishers, Dordrecht, the Netherlands.

Tennekes H. and Lumley J.L. 1972 *A first course in turbulence.* MIT Press, Cambridge, Mass.

Weil, J.C. 1988 Disperion in the convective boundary layer. In *Lectures on air pollution modelling* (eds. A. Venkatram and J.C. Wyngaard), American Meteorological Society, Boston, U.S.A., pp 167–222.

Wyngaard, J.C. 1988 Structure of the PBL. In *Lectures on air pollution modelling* (eds. A. Venkatram and J.C. Wyngaard), American Meteorological Society, Boston, U.S.A., pp 9–61.

Wyngaard, J.C. 1992 *Atmospheric turbulence.* Annu. Rev. Fluid Mech..24205–233

Wyngaard, J.C., Coté O.R. and Y. Izumi 1971 Local free convection, similarity and the budgets of shear stress and heat flux. *J. Atmos. Sci.* 28, 1171–1182.

Printed in the United States
By Bookmasters